Reviews of Environmental Contamination and Toxicology

VOLUME 165

Springer
*New York
Berlin
Heidelberg
Barcelona
Hong Kong
London
Milan
Paris
Singapore
Tokyo*

Reviews of Environmental Contamination and Toxicology

Continuation of Residue Reviews

Editor
George W. Ware

Editorial Board
Lilia A. Albert, Xalapa, Veracruz, Mexico
F. Bro-Rasmussen, Lyngby, Denmark · D.G. Crosby, Davis, California, USA
Pim de Voogt, Amsterdam, The Netherlands · H. Frehse, Leverkusen-Bayerwerk, Germany
O. Hutzinger, Bayreuth, Germany · Foster L. Mayer, Gulf Breeze, Florida, USA
N.N. Melnikov, Moscow, Russia · D.P. Morgan, Cedar Rapids, Iowa, USA
Douglas L. Park, Baton Rouge, Louisiana, USA
Annette E. Pipe, Burnaby, British Columbia, Canada
Raymond S.H. Yang, Fort Collins, Colorado, USA

Founding Editor
Francis A. Gunther

VOLUME 165

Springer

Coordinating Board of Editors

DR. GEORGE W. WARE, *Editor*
Reviews of Environmental Contamination and Toxicology

5794 E. Camino del Celador
Tucson, Arizona 85750, USA
(520) 299-3735 (phone and FAX)

DR. HERBERT N. NIGG, *Editor*
Bulletin of Environmental Contamination and Toxicology

University of Florida
700 Experimental Station Road
Lake Alfred, Florida 33850, USA
(941) 956-1151; FAX (941) 956-4631

DR. DANIEL R. DOERGE, *Editor*
Archives of Environmental Contamination and Toxicology

6022 Southwind Drive
N. Little Rock, Arkansas, 72118, USA
(501) 791-3555; FAX (501) 791-2499

Springer-Verlag
New York: 175 Fifth Avenue, New York, NY 10010, USA
Heidelberg: Postfach 10 52 80, 69042 Heidelberg, Germany

Library of Congress Catalog Card Number 62-18595.
Printed in the United States of America.

ISSN 0179-5953

Printed on acid-free paper.

© 2000 by Springer-Verlag New York, Inc.
All rights reserved. This work may not be translated or copied in whole or in part without the written permission of the publisher (Springer-Verlag New York, Inc., 175 Fifth Avenue, New York, NY 10010, USA), except for brief excerpts in connection with reviews or scholarly analysis. Use in connection with any form of information storage and retrieval, electronic adaptation, computer software, or by similar or dissimilar methodology now known or hereafter developed is forbidden. The use of general descriptive names, trade names, trademarks, etc., in this publication, even if the former are not especially identified, is not to be taken as a sign that such names, as understood by the Trade Marks and Merchandise Marks Act, may accordingly be used freely by anyone.

ISBN 0-387-95013-3 Springer-Verlag New York Berlin Heidelberg   SPIN 10757837

Foreword

International concern in scientific, industrial, and governmental communities over traces of xenobiotics in foods and in both abiotic and biotic environments has justified the present triumvirate of specialized publications in this field: comprehensive reviews, rapidly published research papers and progress reports, and archival documentations. These three international publications are integrated and scheduled to provide the coherency essential for nonduplicative and current progress in a field as dynamic and complex as environmental contamination and toxicology. This series is reserved exclusively for the diversified literature on "toxic" chemicals in our food, our feeds, our homes, recreational and working surroundings, our domestic animals, our wildlife and ourselves. Tremendous efforts worldwide have been mobilized to evaluate the nature, presence, magnitude, fate, and toxicology of the chemicals loosed upon the earth. Among the sequelae of this broad new emphasis is an undeniable need for an articulated set of authoritative publications, where one can find the latest important world literature produced by these emerging areas of science together with documentation of pertinent ancillary legislation.

Research directors and legislative or administrative advisers do not have the time to scan the escalating number of technical publications that may contain articles important to current responsibility. Rather, these individuals need the background provided by detailed reviews and the assurance that the latest information is made available to them, all with minimal literature searching. Similarly, the scientist assigned or attracted to a new problem is required to glean all literature pertinent to the task, to publish new developments or important new experimental details quickly, to inform others of findings that might alter their own efforts, and eventually to publish all his/her supporting data and conclusions for archival purposes.

In the fields of environmental contamination and toxicology, the sum of these concerns and responsibilities is decisively addressed by the uniform, encompassing, and timely publication format of the Springer-Verlag (Heidelberg and New York) triumvirate:

Reviews of Environmental Contamination and Toxicology [Vol. 1 through 97 (1962–1986) as Residue Reviews] for detailed review articles concerned with any aspects of chemical contaminants, including pesticides, in the total environment with toxicological considerations and consequences.

Bulletin of Environmental Contamination and Toxicology (Vol. 1 in 1966) for rapid publication of short reports of significant advances and discoveries in the fields of air, soil, water, and food contamination and pollution as well as methodology and other disciplines concerned with the introduction, presence, and effects of toxicants in the total environment.

Archives of Environmental Contamination and Toxicology (Vol.1 in 1973) for important complete articles emphasizing and describing original experimental or theoretical research work pertaining to the scientific aspects of chemical contaminants in the environment.

Manuscripts for *Reviews* and the *Archives* are in identical formats and are peer reviewed by scientists in the field for adequacy and value; manuscripts for the *Bulletin* are also reviewed, but are published by photo-offset from camera-ready copy to provide the latest results with minimum delay. The individual editors of these three publications comprise the joint Coordinating Board of Editors with referral within the Board of manuscripts submitted to one publication but deemed by major emphasis or length more suitable for one of the others.

Coordinating Board of Editors

Preface

Thanks to our news media, today's lay person may be familiar with such environmental topics as ozone depletion, global warming, greenhouse effect, nuclear and toxic waste disposal, massive marine oil spills, acid rain resulting from atmospheric SO_2 and NO_x, contamination of the marine commons, deforestation, radioactive leaks from nuclear power generators, free chlorine and CFC (chlorofluorocarbon) effects on the ozone layer, mad cow disease, pesticide residues in foods, green chemistry or green technology, volatile organic compounds (VOCs), hormone- or endocrine-disrupting chemicals, declining sperm counts, and immune system suppression by pesticides, just to cite a few. Some of the more current, and perhaps less familiar, additions include *xenobiotic transport, solute transport, Tiers 1 and 2, USEPA to cabinet status, and zero-discharge*. These are only the most prevalent topics of national interest. In more localized settings, residents are faced with leaking underground fuel tanks, movement of nitrates and industrial solvents into groundwater, air pollution and "stay-indoors" alerts in our major cities, radon seepage into homes, poor indoor air quality, chemical spills from overturned railroad tank cars, suspected health effects from living near high-voltage transmission lines, and food contamination by "flesh-eating" bacteria and other fungal or bacterial toxins.

It should then come as no surprise that the '90s generation is the first of mankind to have become afflicted with *chemophobia*, the pervasive and acute fear of chemicals.

There is abundant evidence, however, that virtually all organic chemicals are degraded or dissipated in our not-so-fragile environment, despite efforts by environmental ethicists and the media to persuade us otherwise. However, for most scientists involved in environmental contaminant reduction, there is indeed room for improvement in all spheres.

Environmentalism is the newest global political force, resulting in the emergence of multi-national consortia to control pollution and the evolution of the environmental ethic. Will the new politics of the 21st century be a consortium of technologists and environmentalists or a progressive confrontation? These matters are of genuine concern to governmental agencies and legislative bodies around the world, for many serious chemical incidents have resulted from accidents and improper use.

For those who make the decisions about how our planet is managed, there is an ongoing need for continual surveillance and intelligent controls to avoid endangering the environment, the public health, and wildlife. Ensuring safety-in-use of the many chemicals involved in our highly industrialized culture is a dynamic challenge, for the old, established materials are continually being displaced by newly developed molecules more acceptable to federal and state regulatory agencies, public health officials, and environmentalists.

Adequate safety-in-use evaluations of all chemicals persistent in our air, foodstuffs, and drinking water are not simple matters, and they incorporate the judgments of many individuals highly trained in a variety of complex biological, chemical, food technological, medical, pharmacological, and toxicological disciplines.

Reviews of Environmental Contamination and Toxicology continues to serve as an integrating factor both in focusing attention on those matters requiring further study and in collating for variously trained readers current knowledge in specific important areas involved with chemical contaminants in the total environment. Previous volumes of *Reviews* illustrate these objectives.

Because manuscripts are published in the order in which they are received in final form, it may seem that some important aspects of analytical chemistry, bioaccumulation, biochemistry, human and animal medicine, legislation, pharmacology, physiology, regulation, and toxicology have been neglected at times. However, these apparent omissions are recognized, and pertinent manuscripts are in preparation. The field is so very large and the interests in it are so varied that the Editor and the Editorial Board earnestly solicit authors and suggestions of underrepresented topics to make this international book series yet more useful and worthwhile.

Reviews of Environmental Contamination and Toxicology attempts to provide concise, critical reviews of timely advances, philosophy, and significant areas of accomplished or needed endeavor in the total field of xenobiotics in any segment of the environment, as well as toxicological implications. These reviews can be either general or specific, but properly they may lie in the domains of analytical chemistry and its methodology, biochemistry, human and animal medicine, legislation, pharmacology, physiology, regulation, and toxicology. Certain affairs in food technology concerned specifically with pesticide and other food-additive problems are also appropriate subjects.

Justification for the preparation of any review for this book series is that it deals with some aspect of the many real problems arising from the presence of any foreign chemical in our surroundings. Thus, manuscripts may encompass case studies from any country. Added plant or animal pest-control chemicals or their metabolites that may persist into food and animal feeds are within this scope. Food additives (substances deliberately added to foods for flavor, odor, appearance, and preservation, as well as those inadvertently added during manufacture, packing, distribution, and storage) are also considered suitable review material. Additionally, chemical contamination in any manner of air, water, soil, or plant or animal life is within these objectives and their purview.

Normally, manuscripts are contributed by invitation, but suggested topics are welcome. Preliminary communication with the Editor is recommended before volunteered review manuscripts are submitted.

Tucson, Arizona

G.W.W.

Table of Contents

Foreword v
Preface vii

Mesocosms in Ecotoxicology (1): Outdoor Aquatic Systems 1
    THIERRY CAQUET, LAURENT LAGADIC, and STEVEN R. SHEFFIELD

Lizard Contaminant Data for Ecological Risk Assessment 39
    KYM ROUSE CAMPBELL and TODD S. CAMBELL

Biomarkers in Earthworms 117
    JANECK J. SCOTT-FORDSMAND and JASON M. WEEKS

Index 161

Mesocosms in Ecotoxicology (1): Outdoor Aquatic Systems

Thierry Caquet, Laurent Lagadic, and Steven R. Sheffield

Contents

I. Introduction	1
II. General Considerations	2
A. Definitions	2
B. Design/Construction Variables	3
III. Current Practices	6
A. Initial Composition	6
B. Stabilization Period	11
C. Implementation of Testing Procedures and Exposure to Chemicals	12
IV. Future Developments	16
A. Predictive Role of Mesocosms	16
B. Use of Mesocosms for Ecotoxicological Risk Assessment	18
V. Conclusions	20
Summary	21
Acknowledgments	22
References	23

I. Introduction

The first deliberately constructed artificial aquatic ecosystems were designed for ecological studies to develop and validate new theories on ecosystem structure and function (Hall et al. 1970; Lawton 1995). Such systems are currently used worldwide and seem to be promising in the developing field of ecological engineering (Kangas and Adey 1996; Odum 1996). It became rapidly evident that model ecosystems could also provide valuable information for the assessment of the fate and effects of chemicals. In particular, they provide the opportunity to simultaneously identify direct and indirect effects of toxicants and to investigate

---

Communicated by George W. Ware.

T. Caquet·L. Lagadic (✉)
INRA, Equipe d'Ecotoxicologie Aquatique, Station Commune de Recherche en Ichtyophysiologie, Biodiversité et Environnement, Campus de Beaulieu, F-35042 Rennes Cedex, France.
T. Caquet
Laboratoire d'Ecologie et de Zoologie, UPRESA CNRS 8079, Bât. 442, Université de Paris-Sud, F-91405 Orsay Cedex, France.
S.R. Sheffield
The Institute of Wildlife and Environmental Toxicology, Department of Environmental Toxicology, Clemson University, P.O. Box 709, Pendleton, SC 29670, U.S.A.
Current address: U.S. Fish and Wildlife Service, 4401 N. Fairfax Drive, Suite 634, Arlington, VA 22203, U.S.A. steven_r_sheffield@fws.gov

responses at many different levels of biological organization in fairly controlled conditions of exposure. Processes that reduce, such as adsorption on suspended solids or sediments, or enhance, such as bioturbation or bioaccumulation, the bioavailability of contaminants can also be taken into account using such experimental devices. Because of their relevance in environmental risk assessment, mesocosms have sometimes been required for the registration of new chemicals, especially pesticides, and corresponding guidelines and guidance documents have been proposed (Touart 1988; Crossland 1990; SETAC-RESOLVE 1992; SETAC-Europe 1992). New guidelines are currently under evaluation (OECD 1996; USEPA-OPPTS 1996).

Different approaches in physically confined, multitrophic, and self-maintaining indoor and outdoor systems have been developed for ecotoxicity testing of chemicals (Beyers and Odum 1993; Cairns and Niederlehner 1994; Kennedy et al. 1995; Cairns et al. 1996). The present review is concerned only with outdoor systems. The originality of outdoor microcosms and mesocosms is based mainly on the combination of ecological realism, achieved by introduction of the basic components of natural ecosystems, and facilitated access to a number of physicochemical, biological, and toxicological parameters that can be controlled to some extent. In other words, chemical testing in such systems is more realistic than laboratory tests and easier than field assessment of chemical effects (Odum 1984; Cairns 1988; Crossland 1994; La Point 1994). In this respect, it should be mentioned that the more suitable approach for using model ecosystems in ecotoxicological studies consists in coupling with standardized laboratory tests.

Basic features of mesocosms have already been thoroughly reviewed (Crossland and Bennett 1989; Kosinski 1989; Ravera 1989; Guckert 1993; Belanger 1994, 1997; Brock and Budde 1994; Graney et al. 1994, 1995; Heimbach 1994; Hill et al. 1994a; Mitchell 1994; Kennedy et al. 1995; Rodgers et al. 1996). In the following review, emphasis is placed on structural and methodological aspects that can be critical for the use of outdoor mesocosms in environmental risk assessment of chemicals.

II. General Considerations
A. Definitions

Broadly speaking, mesocosms are man-made, surrogate ecosystems that can be used to assess the fate and effects of chemicals at many different levels of biological organization through appropriate endpoints. According to the classical definition of Odum (1984), mesocosms are "bounded and partially enclosed outdoor experimental setups . . . falling between laboratory microcosms and the large, complex, real world macrocosms." Since this original paper, the concept has been refined and mesocosms are frequently referred to as experimental or model ecosystems, i.e., "physically confined multitrophic and self-maintaining systems . . . which have a duration time exceeding the penultimate trophic level present and whose size is sufficient to enable pertinent sampling and measurements to be made without seriously influencing the structure and dynamics of the system" (Lalli 1990). More recently, Graney et al. (1995) proposed the use of "simulated field studies" to describe experiments performed in devices that correspond to either isolated subsections of the natural environment or manmade physical models of lotic or lentic ecosystems. In fact, the term mesocosm is currently used to describe indoor and outdoor artificial streams or experimental ponds and enclosures (Fig. 1).

Among enclosures, a distinction can be made between limnocorrals, pelagic bags, and littoral enclosures. Limnocorrals are placed in the pelagic region of ponds, lakes, or marine environments; they may or may not be in contact with the bottom sediments. Pelagic bags are a particular type of limnocorrals that enclose only the water column. Littoral enclosures are built using dividers that isolate the littoral region (maximum depth, 2–4 m) of ponds or lakes. These systems have been frequently used to assess the fate of chemicals in the aquatic environment and the impact of pesticides or other toxicants on planktonic and macroinvertebrate communities and sometimes fish, either in the freshwater (Shires 1983; Stephenson and Kane 1984; Solomon et al. 1985, 1986, 1989; Kaushik et al. 1985, 1986; Herman et al. 1986; Stephenson et al. 1986; Day et al. 1987; Brazner et al. 1989; Hamilton et al. 1989; Hanazato et al. 1989; Havens and Heath 1989; Siefert et al. 1989; Brazner and Kline 1990; Wangberg et al. 1991; Wayland 1991; Liber et al. 1992, 1997; Lozano et al. 1992; Lay et al. 1993; Havens 1995; Kreutzweiser and Thomas 1995; Tanner and Knuth 1995; O'Halloran et al. 1996; Jak et al. 1998; Sierszen and Lozano 1998) or in the marine or estuarine environment (Menzel and Case 1977; Sörensson et al. 1989; Lalli 1990; Clark and Noles 1994). Experimental ditches, extensively used in The Netherlands (Drent and Kersting 1993; Crum and Brock 1994; Verdonschot and Ter Braak 1994; van den Brink et al. 1996; van Wijngaarden et al. 1996; Verdonschot 1996; Kersting and van der Brink 1997; Ronday et al. 1998), are a particular type of mesocosm since they can be used as either static or flow-through systems (Cid Montañés et al. 1995).

B. Design/Construction Variables

Size Versus Self-Sustainability

Streams and ponds of various sizes have been used, ranging from less than 1 m to 520 m in length for streams and from 2 to 1000 m^3 in volume for ponds; limnocorrals vary from 2 L to more than 2.5×10^6 L (generally between 1,000 and 10,000 L). Mesocosm size (surface and volume for ponds and enclosures, length for streams) has frequently been identified as a criterion to distinguish between microcosms and mesocosms (SETAC-Europe 1992; Heimbach 1994). This distinction points out the importance of a minimal size for aquatic mesocosms. In the literature, the term mesocosm actually refers to model systems that largely differ in size, e.g., from a few to several hundred cubic meters for pond mesocosms, and long-term (several months to a few years) self-sustainability rather appears as the major common feature of those artificial ecosystems (Voshell 1990). Therefore, the minimal size of an aquatic mesocosm may be

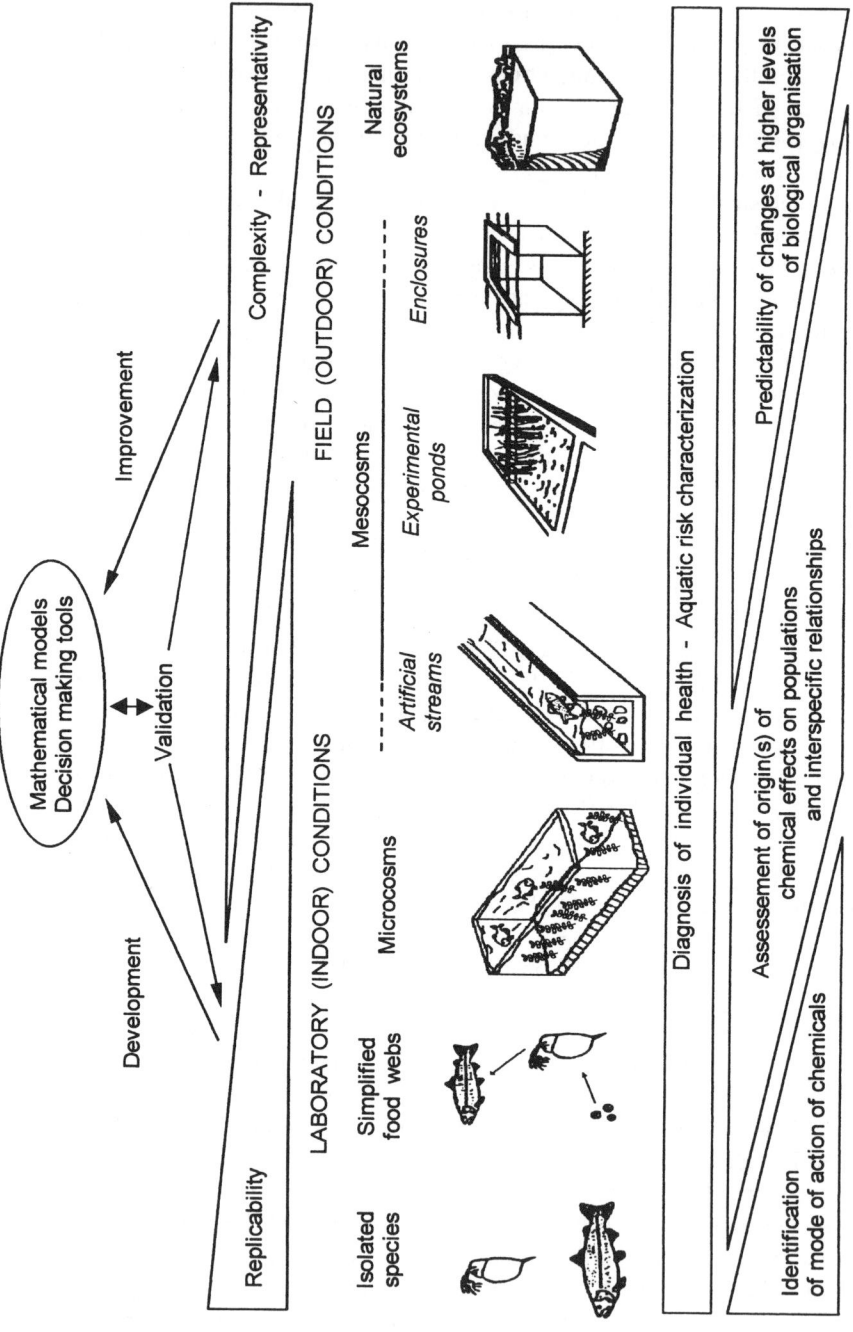

Fig. 1. Place of mesocosms among the various experimental contexts used in aquatic ecotoxicology (see text for explanation). (Modified from Caquet et al. 1996 and Lagadic 1999.)

that which is sufficient to maintain appropriate proportions of all the components that are necessary for long-term self-functioning of the ecosystem or that which is necessary to focus on specific endpoints. It is necessary that the experimental systems have sufficient complexity in their functional attributes so that direct and indirect interactions, e.g., intra- and interspecific competition for resources and predation, between all major functional groups can occur. On the other hand, the systems should be of sufficient size so that sampling during experiments does not have a major impact. The maximal size of mesocosms mainly refers to structural and functional homogeneity, and also to control of the conditions of chemical application (La Point 1994). One of the main pitfalls of the largest lentic systems (several hundred acres) is the risk of structural and functional divergence between different parts of the same mesocosm, thus resulting in increased variability of measurements and lack of replicability. More recent studies tend to use rather size-limited systems, 10–20 m^3, or 10–20 m long, for outdoor pond or stream mesocosms, respectively (Crossland 1994; Matthiessen 1994; Caquet et al. 1996; Shaw and Kennedy 1996; Shaw and Manning 1996; Belanger 1997). Because sampling in any compartment does not have a major impact on the structure and functioning of the systems, these dimensions may represent appropriate sizes for mesocosms that could be used for regulatory purposes.

Importance of Shape

The shape of the edges of mesocosms determines the importance of the so-called wall-effects, which may greatly reduce the interest of studies performed in such systems. For example, the vertical walls of enclosures result in an important development of fouling organisms such as periphyton, both inside and outside the system, which in turn results in changes in the abundance and structure of the invertebrate and vertebrate communities as well as in various physico-chemical parameters such as nutrient concentration, turbulence, temperature, or light (Stephenson et al. 1984; Arumugam and Geddes 1986; Chant and Cornett 1988; Siefert et al. 1990; Heinis and Knuth 1992). Therefore, whenever possible the shape of the edges should be designed to reduce the wall-effects, for example, by using moderately sloped walls (Layton and Voshell 1991; Heimbach et al. 1994; Howick et al. 1994; Rosenzweig and Buikema 1994; Jenkins 1995; Caquet et al. 1996). When wall-effects are expected, they must be taken into account in sampling design (Stephenson et al. 1984, 1986; Liber 1994).

Artificial streams are frequently divided into successive subsystems differing in water depth and velocity and sometimes in bed characteristics. Although several patterns have been proposed and used, the most common feature for these systems consists of successive pools and riffles that exhibit contrasting characteristics from both biotic and abiotic aspects (Pignatello et al. 1983; Fairchild et al. 1987; Arthur 1988; Allen 1991; Crossland et al. 1991, 1992; Mitchell et al. 1993; Swift et al. 1993; Rodgers et al. 1996; Dorn et al. 1997; Belanger 1997).

Importance of Materials Used

Whether dug out or on the ground, building of mesocosms mainly involves two procedures. Hard materials such as concrete or fiberglass, or instead a soft substrate, earth-based mesocosms lined with plastic material, may be used. In any case, the risks of release of compounds from the lining material and of adsorption and possible further release of the tested chemicals should be evaluated. Then, the choice should be directed toward materials that reduce those risks. Similar considerations should be applied to every device that must be introduced in the water for sampling or organism maintenance.

III. Current Practices
A. Initial Composition

The initial composition of mesocosms closely depends on the objectives of the study through the choice of the organisms under testing. However, beside test organisms, other organisms must be introduced because they play key roles in the structure and functioning of the mesocosms.

Basic Requirements

Biophysical Compartments: Water and Sediments

Water. Water can come from 'unpolluted' natural sources (surface or groundwater) or from city sources. Artificial streams generally receive water derived from adjacent natural lotic (Arthur et al. 1983; Pignatello et al. 1983; Lynch et al. 1985; Clements et al. 1988; O'Neil et al. 1990; Crossland et al. 1991, 1992; Hermanutz et al. 1992; Maltby 1992; Swift et al. 1993; Belanger 1994; Pusey et al. 1994a,b; Ward et al. 1995) or lentic ecosystems (Rodgers et al. 1996); more rarely, well water is used (Bowling et al. 1980; Fairchild et al. 1987; Hermanutz et al. 1992). In some cases, streams may receive polluted water derived from an adjacent watercourse (Bankey et al. 1995; Hatakeyama et al. 1997). When water derived from a natural stream is used, the seasonal abundance of suspended solids should be carefully assessed because flow velocity is generally lower in artificial than in natural streams and significant sedimentation may occur in the former. Furthermore, spates or heavy rainfall may dramatically influence water characteristics within the experimental systems, as through rapid changes in turbidity (Cuffney et al. 1990; Crossland et al. 1991; Rodgers et al. 1996). Pond mesocosms are filled with either well or groundwater (Ali and Stanley 1982; Hanazato et al. 1989; Hanazato and Yasuno 1990a,b; Heimbach et al. 1992; Cid Montañés et al. 1995; Kasai and Hanazato 1995a,b), tap water (Caquet et al. 1992, 1996; Jenkins 1995), or water from a lake, reservoir, or pond (Hurlbert et al. 1970, 1972; deNoyelles et al. 1982; Shires 1983; Giddings et al. 1984; Kaushik et al. 1985; Solomon et al. 1985, 1989; Dewey 1986; Stephenson and Mackie 1986; Stephenson et al. 1986; Kettle et al. 1987; Threlkeld and Soballe 1988; Yasuno et al. 1988; Brazner et al. 1989; Maguire et al.

1989; Brazner and Kline 1990; Siefert et al. 1990; Wayland and Boag 1990; Heinis and Knuth 1992; Liber et al. 1992; Lozano et al. 1992; Getsinger et al. 1994). Natural surface waters do not allow reliable control of the organisms, mainly plankton, that will eventually be introduced in the mesocosms, and this source needs to be thoroughly investigated before testing of chemicals. City water usually undergoes chlorination and sometimes fluoridation, and therefore needs to be dechlorinated or defluoridated before introduction of organisms into the mesocosms. As neither natural nor city water may be free of pollutants, chemical analysis should be performed to quantify levels of potential contaminants. Introduced water should also be analyzed for the classical physicochemical parameters (pH, hardness, turbidity, conductivity, dissolved oxygen, nitrogen and ammonium, phosphorus, etc.). Whether fixed or season variable, which is more ecologically realistic, water level should always be sufficient to allow the development and maintenance of groups of organisms that are necessary for a convenient structure and functioning of the mesocosms. Adding water to maintain a constant water level during the period of chemical testing results in both dilution of compounds, both parent compound and metabolites, and introduction of contaminants (chlorides, pesticides, and other pollutants) and organisms that can exert confounding effects.

Sediments. Current guidelines recommend the use of natural sediments from "uncontaminated" reference ecosystems (SETAC-Europe 1992; OECD 1996; USEPA-OPPTS 1996). They also recommend structural and chemical analysis, and mixing before introduction into the mesocosms. The main advantage of such sediments is obviously their inherent physical and biological properties that render mesocosms rapidly functional (Hanazato and Yasuno 1987, 1990a,b; Ferrington et al. 1994; Kasai and Hanazato 1995b). Some studies have been performed using topsoil as the sediment (Stephenson and Mackie 1986; Layton and Voshell 1991; Rosenzweig and Buikema 1994; Sparling and Lowe 1996), but this approach presents various disadvantages, e.g., lack of functionality and increase in maturation delay. Another form of natural sediment, contaminated dredge spoil, has been used successfully as sediment in pond mesocosms in order to examine biomarkers of health in fishes (Vethaak et al. 1996). In stream mesocosms, sediments consist frequently of trays filled with pebbles and cobbles that have been left in natural streams for several weeks or months (Clements et al. 1989, 1990; Lowe et al. 1996). These trays may be used to sample benthic organisms during the experiment (Clements et al. 1989; Lowe et al. 1996).

The use of natural sediments ensures the establishment of communities of primary producers and, to a lesser extent, consumers through the resting stages in the sediments. Sediment depth should be calculated relative to the size of the mesocosms and hence to water depth. A sediment-to-water ratio of 10% can be considered as appropriate for pond mesocosms. When natural sediments are introduced in the mesocosms, care should be taken to ensure that the relative abundance of the various sediment granulometric fractions is the same in the different replicates. For example, in artificial streams, fine particles (<1-mm diameter) tend to fill the interstitial spaces in the gravel, thus reducing available habitat, which in turn reduces the diversity and abundance of invertebrates (Crossland et al. 1992). The optimal percentage of fine particles seems to be 12%–14% (Adams and Beschta 1980; Fairchild et al. 1987). Granulometry of sediments plays an important role in the colonization of lentic systems by insects such as chironomids (Francis and Kane 1995). It is also of major importance for fish, because sediments provide foraging substrates and sites for spawning and egg deposition (Resetarits 1991; Gelwick and Matthews 1993).

The use of natural sediments requires that suitable noncontaminated substrates with similar abiotic and biotic characteristics are available at any time. Important disadvantages of using natural sediments arise when the number of mesocosms under testing is important because dozens of cubic meters are generally necessary to build operational systems. Beside technical and financial aspects, the transfer of natural sediments to the mesocosms may result in important physical and chemical changes. Thus, destructuration of sediment layers or sudden release of organic or inorganic compounds (e.g., nutrients) may result in further biological discrepancies between mesocosms. Moreover, if natural sediments contain organisms that are necessary to obtain ecologically representative systems, they can also include undesirable hosts such as plant seeds or parasitic animals which may cause trouble in further experiments, especially in shallow artificial systems.

Artificial substrates appear to be an attractive opportunity to overcome most of the drawbacks of natural sediments. Artificial sediments have tentatively been proposed for laboratory toxicity testing (Burton 1996), and their composition could be used as a rational basis for a definition of artificial sediments to be introduced in mesocosms. However, artificial substrates undoubtedly lack a well-established functional microbial community and therefore cannot be used for long-term experiments. The use of mixtures of natural and artificial sediments may constitute an interesting alternative because this would ensure reproducible physicochemical composition and initial biological activity.

Structural and Functional Groups of Organisms

In addition to the organisms present in sediments and water from natural sources, introduction of organisms that render mesocosms structurally and functionally suitable for chemical testing is usually preferable.

Structural groups of organisms are mainly aquatic vascular plants or macrophytes. Among other roles, macrophytes provide substrates for the development of food resources (periphyton) or for egg laying and development of juvenile stages of invertebrates; they also provide refuges for animals such as invertebrates or fish, especially at juvenile stages, from predators (Gilinsky 1984; Schramm et al. 1987; Learner et al. 1989; Schramm and Jirka 1989; Newman 1991; Jeffries 1993; Blake 1994). They may also interfere with colonization processes (Angerilli and Beirne 1980; Jeffries 1993). Other groups of organisms can also play a significant role in the structuration of mesocosms. For example, the vertical movements of plankton or the presence of predators can influence the spatial distribution of plankton feeders or prey, respectively. Functional groups of aquatic organisms are mainly primary producers (phytoplankton and, to a lesser extent, macrophytes) and decomposers (microorganisms and invertebrates). These two communities play a fundamental role in energy and nutrient flow and cycling in such systems. Primary and secondary consumers are also of importance because their abundance reflects the productivity and welfare of the systems (Brock and Budde 1994; Kersting 1994; La Point 1994).

The introduction of macroinvertebrates in mesocosms may be achieved either by sampling in natural ecosystems (deNoyelles et al. 1982; Dewey 1986; Wayland and Boag 1990; Caquet et al. 1992, 1996; Crossland et al. 1992; Maltby 1992; Baturo et al. 1995; Cid Montañés et al. 1995), sometimes with the substrate (Lynch et al. 1985), or by introducing artificial substrates that have previously been placed in situ for several days or weeks for colonization (Clements et al. 1988, 1989, 1990; O'Neil et al. 1990; Crossland et al. 1992; Lowe et al. 1996). Some organisms can also be introduced with water from natural sources, especially in artificial streams (Hermanutz et al. 1992; Pusey et al. 1994a,b; Ward et al. 1995). The introduction of invertebrates can be repeated many times during the stabilization period (Crossland et al. 1992). Spontaneous colonization of lentic mesocosms by macroinvertebrates, e.g., insects, also take place but requires several months (Street and Titmus 1979; Layton and Voshell 1991). When natural colonization is the only mechanism involved, aquatic communities that develop in lentic mesocosms are dominated by detritus-feeding organisms, e.g., chironomids, and to a lesser extent by flying predators, e.g., Coleoptera and Heteroptera, a common feature in colonization dynamics of aquatic ecosystems (Paterson and Fernando 1970; Street and Titmus 1979; Barnes 1983; Voshell and Simmons 1984; Christman and Voshell 1993).

The introduction of fish in experimental systems raises many questions about the choice of the species or group of species, the loading rates, and the endpoints that can be measured (Fairchild and Little 1993). Fish are known to have direct and indirect effects on ecosystem functioning. They can significantly alter both phyto- and zooplankton community structure and function and nutrient dynamics and cycling either through direct (e.g., selective predation) or indirect (e.g., reduced grazing pressure) effects (Brooks and Dodson 1965; Andersson et al. 1978; Carpenter et al. 1985; Drenner et al. 1986; Brabrand et al. 1987; Langeland et al. 1987; Mazumder et al. 1988; Vinyard et al. 1988; Reinertsen et al. 1990; Lazzaro et al. 1992; Qin and Culver 1996; Vanni and Layne 1997; Vanni et al. 1997). Proposed guidelines (Touart 1988; USEPA-OPPTS 1996) require that mesocosms include a reproductive population of bluegill sunfish (*Lepomis macrochirus*). However, several studies demonstrated that the presence of these animals obscured or complicated the evaluation of pesticide impacts on invertebrate populations (Fairchild et al. 1992; Howick et al. 1992, 1994; Weber et al. 1992; Mayasich et al. 1994; Morris et al. 1994; Kennedy et al. 1995), although piscivorous species such as largemouth bass (*Micropterus salmoides*) may be used to control planktivorous or insectivorous fishes such as bluegill (Deutsch et al. 1992; Howick et al. 1993). If fish must be introduced in mesocosms, two different approaches should be considered. In the first approach, the fish only contribute to the ecological realism of mesocosms, and the choice of the species is therefore mainly driven by the effects they may have on system structure and functioning. In the second approach, fish are considered as a model for ecotoxicological studies. The species chosen should therefore be able to reproduce in the mesocosm; it also should be easily sampled, and the impact of the offspring should not induce dramatic changes in mesocosm structure and functioning. In this case, species with a short reproductive cycle, such as fathead minnow (*Pimephales promelas*) and three-spined stickleback (*Gasterosteus aculeatus*), have been proposed for long-term investigations (OECD 1996). Several studies have already demonstrated that these species are able to hatch, develop, grow, and even reproduce in adequately designed systems (Lehtinen 1989; Shaw et al. 1995a,b,c; Dorn et al. 1996, 1997; Kline et al. 1996; Harrelson et al. 1997). The number of fish initially introduced in mesocosms depends on the approach, on biological characteristics of the species, and on the size of the experimental systems.

Organisms to Be Tested

The various constraints of time, money, staff, and ease of sampling require that some choices should be made concerning the organisms on which the effects of toxicants are to be assessed. This approach is much more realistic than that previously promoted by Touart (1988), who proposed to "measure everything" within the mesocosms.

Choice. Apart from the foregoing considerations specific to fish, the choice of organisms placed in mesocosms to be tested for the effects of a given chemical closely depends on the objectives of the study. Preference is generally given (1) to species for which a certain amount of (eco)toxicological knowledge already exists, and (2) to target species that are likely to encounter toxic effects in their natural environment.

Origin. The organisms introduced in mesocosms may come from laboratory breeding strains (Taylor et al. 1994) or from field-sampled organisms (deNoyelles et al. 1982; Clements et al. 1988, 1989; O'Neil et al. 1990; Wayland and Boag 1990; Caquet et al. 1992, 1996; Crossland et al. 1992; Maltby 1992; Baturo et al. 1995; Baturo and Lagadic 1996). The main advantages of laboratory organisms are a fairly good knowledge of their genetic and biological characteristics and the availability of large numbers of individuals of known age or developmental stage. The main problem with using laboratory-reared organisms is related to their ability to undergo the stress of transfer from laboratory conditions to outdoor conditions.

Maintenance. Test organisms may be introduced as free individuals or as caged specimens. The former play a significant role in mesocosm structure and functioning, but sampling is rather difficult and may constitute an important source of perturbation. Caging facilitates sampling of test species and can also favor contacts between organisms and the tested chemical at a particular location in the mesocosms. However, caging may significantly alter the characteristics of enclosed animals through stress or abnormal environmental conditions, such as elevated temperature. This phenomenon may reduce the interest of such studies unless the effects of caging have correctly been estimated through comparison of selected endpoints with free-ranging individuals of the same species. Even if some developmental stages such as eggs or larval stages are less sensitive to caging, not all species can afford confinement, and this should be clearly evaluated, especially for species that may be caged in the environment.

B. Stabilization Period

Ecological maturity of mesocosms affects the degree of variability of both physicochemical and biological parameters used to investigate the impact of contaminants. It is not always possible to attain equilibrium, whatever definition is given to this term, but adequate time is required to establish a number of interacting functional groups (Giesy and Odum 1980). Studies using limnocorrals and littoral enclosures frequently have no acclimation period because these systems enclose established communities (Solomon et al. 1985; Chant and Cornett 1988; Yasuno et al. 1988; Heinis and Knuth 1992; Lozano et al. 1992). In stream mesocosms, stabilization periods of 10 d (Genter et al. 1987), 4 wk (Crossland et al. 1992), or 1 yr (Lynch et al. 1985) have been reported. The duration of the maturation period for pond mesocosms varies from 1–2 mo (SETAC-RESOLVE 1992) to 2 yr (Hall et al. 1970). According to Rosenzweig and Buikema (1994), a 1-yr stabilization period may be too short to use newly created ponds as replicate test systems. However, their observations, performed on phytoplankton, should be considered with caution because they did not introduce any organisms and used topsoil as sediment.

Following initial system preparation, acclimation usually allows the various biotic components to adjust to the new environment and establish interspecific and abiotic interactions. Duration of the acclimation period depends on both system size and complexity and on study goals (Heimbach 1994; Kennedy et al. 1995). The time needed to equilibrate will increase with initial system complexity, although the use of natural sediments usually shortens duration of the stabilization period because natural maturation processes are enhanced (Kennedy et al. 1995). During the acclimation period, natural processes such as colonization by insects or development of aquatic plants increase system realism but also favor system divergence. This phenomenon is more important for small-scale systems (Cuffney et al. 1990) and is likely to increase with increasing development time (Schüürman 1998). Circulation of water between the different systems has frequently been proposed as a way to limit among-systems variability during this period (Crossland 1984; Crossland and Bennett 1984; Crossland and Wolff 1985; Wolff and Crossland 1985; Crossland et al. 1986; Heimbach et al. 1992; USEPA-OPPTS 1996). In any case, a detailed survey of the pretreatment period should help to evaluate the variability of various descriptors and identify parameters that will be useful to detect treatment effects (Christman et al. 1994).

C. Implementation of Testing Procedures and Exposure to Chemicals

Experimental Design: Statistical Considerations
Choice of the experimental design is of primary importance because of its consequences for both the informative value and the financial cost of mesocosm studies. It appears that the experimental design should be chosen on the basis of the objectives of the study rather than on theoretical considerations (Graney et al. 1989; Regal and Lozano 1994). The design will influence how the results of the study can be interpreted and how the treatment concentrations are established. Three experimental designs are generally used: hypothesis testing, point estimate tests (i.e., concentration–response relationships), and hybrid tests that combine elements of both strategies (Regal and Lozano 1994; Touart 1994; Thompson et al. 1994).

For testing of hypotheses, analysis of variance (ANOVA) and other parametric statistical tests are most often used (Shaw et al. 1994; Stunkard 1994; Kennedy et al. 1999), sometimes in combination with descriptive statistics such as coefficients of variation and diversity and similarity indices. This strategy requires using at least three replicates for both control and treated mesocosms to take the natural variability into account. However, it has been suggested that, in most cases, even more replicates would be unable to detect certain ecosystem-level effects or effects on rare taxa because of the high variance values (Rodgers 1994; Kennedy et al. 1995). Moreover, violation of statistical assumptions, e.g., normal distribution of the data or homogeneity of the variances, is a frequent feature of such experiments that may lead to erroneous conclusions (Kennedy et al. 1999). On the other hand, the use of nonparametric statistics would result in a need to increase the number of replicates (Rodgers 1994). The use of the log-linear regression models, which can be used even with zero counts, has recently been proposed as a possible alternative to common statistical methods, and this approach seems very promising (Ammann et al. 1997; Kennedy et al. 1999). Multivariate analysis can also be applied to the data obtained in mesocosms, especially through the use of similarity–dissimilarity indices of communities of aquatic organisms (Clarke 1999; Kedwards et al. 1999; Sparks et al. 1999; van den Brink and Ter Braak 1999; Whittle et al. 1999). In this case, comparison of the dynamics of the communities in control and treated mesocosms is highly relevant.

Point-estimate tests are generally conducted like a classical concentration–response study. The aim of this approach is to determine the shape of the concentration–response relationship for selected criteria and sometimes a NOEC. An ideal design should include both a no-effect concentration and a very effective one, with at least one intermediate effect concentration. This design assumes that the relationship between concentration and effect is monotonous, either increasing or decreasing. Data from such experiments can be analyzed using more or less sophisticated regression models depending on the data (Graney et al. 1995). It is generally assumed that this design does not necessitate the replication of treatment, although replicates, especially for control mesocosms, significantly increase the value of the experiments.

A hybrid approach including both replication of treatments and concentration–response design has been suggested in guidelines proposed for a harmonized use of mesocosms (Touart 1988; US-EPA OPPTS 1996). Graney et al. (1989) indicated that the probability of detecting significant effects at a low concentration, such as estimated environmental concentration, was greater for this approach than for the ANOVA approach.

Whatever the experimental design, the intrinsic variability of the endpoints should be assessed before or during the measurement period. Identification of contaminant effects for highly variable endpoints will generally be more difficult and necessitate more replicates than for endpoints that are less variable or for which a precise pattern of natural variation has already been identified. Common statistical methods use threshold levels of 1% to 5% for rejection of the null hypothesis (type I error; i.e., the risk to reject the hypothesis that treatment has no effect when it is true). However, the main point in environmental risk assessment is to ensure that type II error (i.e., the risk to accept the hypothesis that treatment has no effect when it is false) is minimized so that power of the test is maximized. Minimization of type II error for threshold levels of 1% to 5% may theoretically be obtained by increasing the number of replicates. However, this may be incompatible with the available human and financial resources and with the variability of measured endpoints (Shaw et al. 1994). Therefore, higher threshold levels, generally 20% but ranging from 15–50%, have been proposed in the case of whole-system ecotoxicological experiments (Southwood 1978; Touart 1988; Christman et al. 1994; Shaw et al. 1994; Stunkard 1994).

Modes of Application

Test chemicals such as pesticides and other toxicants are commonly applied to treatment mesocosms, with application method, frequency, and concentration of test chemical used being the major considerations. The method used for application of the test chemical can have considerable influence on its fate and the subsequent exposure of organisms. It should be chosen in accordance with the objectives of the study and the expected or actual use and behavior of the tested compound in the natural environment. If the experiment is designed to assess the effects of runoff pesticides in artificial streams, a single short-term pulse can be used (Pusey et al. 1994b). However, the exposure of stream biota to a single pulse of pesticide or other xenobiotics is not a common event, and diffuse chronic inputs and repeated contamination are more likely to happen in natural ecosystems. Chronic introduction of toxicants in artificial streams is therefore preferred (Kosinski 1989; Allen 1991; Crossland et al. 1992; Hermanutz et al. 1992; Guckert 1993; Mitchell et al. 1993; Ward et al. 1995; Detenbeck et al.

1996), especially for surfactants, which have been intensively studied in such systems (Fairchild et al. 1993; Belanger 1994; Takamura 1995; Dorn et al. 1996, 1997; Gillespie et al. 1996, 1997; Kline et al. 1996; Harrelson et al. 1997).

In ponds and enclosures, application may be unique or repeated. A single application is used when the acute effects of chemicals are to be evaluated, for example, to assess the effects of accidental discharges or voluntary contamination (worst case studies) (Caquet et al. 1992; Getsinger et al. 1994; Groenendijk et al. 1994; Perschbacher et al. 1996). Repeated applications mimic the chronic exposure of natural ecosystems to low levels of toxicants caused by spray drift or runoff of agrochemicals, for example (de Noyelles et al. 1989; Fairchild et al. 1992; Hoagland et al. 1993; Hamer et al. 1994; Hill et al., 1994b; Kennedy et al. 1994; Farmer et al. 1995; Boyle et al. 1996; Giddings et al. 1996; Perschbacher et al. 1996; Ronday et al. 1998).

Chemical inputs can be determined using contaminant fate models that predict the expected contaminant concentration in natural ecosystems. These models include the pesticide drift model (Ganzelmeier 1997), pesticide surface runoff simulation models such as SWRRB (Simulator for Water Resources in Rural Basins; Arnold et al. 1990), PRZM (Pesticide Root Zone Model; Carsel et al. 1984), or EPIC (Erosion-Productivity Impact Calculator; Williams et al. 1990), and more generalist models such as EXAMS, EXAMS II (Exposure Analysis Modeling System; Burns et al. 1982; Burns and Cline 1985; Burns 1990), or QWASI (Quantitative Water Air Sediment Interaction; Mackay et al. 1983; Mackay 1989, 1990). Initial toxicant concentration can also be determined using values derived from standardized laboratory toxicity tests or field observations of actual residue concentrations (Stay and Jarvinen 1995; Ward et al. 1995).

Dosing methods for mesocosm studies vary with the purpose of the study. The contaminant may be added to the water surface, subsurface, or on the sediments by pouring the active ingredient or a mixture of soil and toxicant (Cushman and Goyert 1984; Oviatt et al. 1987; Wayland 1991; Giddings et al. 1997), spraying (Crossland 1982; Shires 1983; Sugiura et al. 1984; Kaushik et al. 1986; Brazner and Kline 1990; Caquet et al. 1992; Hill et al. 1994b; Cid Montañés et al. 1995; Kreutzweiser and Thomas 1995; O'Halloran et al. 1996; Ronday et al. 1998), pumping (Farke et al. 1985; Zischke et al. 1985; Bakke et al. 1988; Clements et al. 1989; Allen 1991; Crossland et al. 1991), or injection (Wakeham et al. 1983; Crossland and Bennett 1984; Stephenson and Kane 1984; deNoyelles et al. 1989; Havens 1995).

Although most studies using mesocosms have been performed on isolated toxicants, these systems can also be used to assess the fate and effects of mixtures of compounds, especially pesticides (Fairchild et al. 1994). However, this complicates considerably design of the experiments.

Another less commonly used approach for introduction of a test chemical (or chemical mixture) into treatment mesocosms is the use of water and/or sediment already containing the chemical(s). This approach does not require application and exposure to the test chemical(s) would be continuous. This approach was used successfully by Vethaak et al. (1996) in their study of biomarkers of health in fish. They established two mesocosms with clean sand for sediment and a third with contaminated dredge spoils. The contaminated dredge spoils mesocosm shared common water circulation with one of the clean sand mesocosms, thereby allowing for indirect contamination via the water phase in one of their mesocosms.

Duration of the Experiment

Duration of the experimental period is closely related to the objective of the study, the behavior of the tested chemicals, and the type of experimental system. Enclosures and streams are generally used for short- or medium-term studies, from several weeks to a few months, whereas ponds can be used for several months or years. This difference mainly arises from structural and functional differences between systems.

Scaling from a spatial and a temporal point of view is also a problem of concern. The choice of appropriate time scales must be considered in the selection of both study duration and sampling frequency intervals between sampling dates. Sampling intervals should consider the life cycles and phenology of important species, the temporal behavior of key physicochemical processes (toxicant half-life, for example), and the longevity of the experimental system (Kennedy et al. 1995). Experimental streams isolated from upstream inputs and riparian ecosystems cannot sustain their own productivity. Moreover, compensatory phenomena of energy and matter drifting, such as upstream movement of animals, is unlikely to occur. Thus, artificial stream community diversity and productivity frequently decrease with time. In enclosures, reduced turbulence and isolation from the rest of the ecosystem lead to nutrient deficiency and sinking of planktonic organisms. The discrepancy between the enclosed area and the surrounding ecosystem thus rapidly increases. Artificial circulation or mixing of water between the bottom and top of the experimental system was frequently used to limit the impact of wall effect on these parameters, especially in pelagic bags (Ravera 1989), but such interventions significantly reduce the representativity of results when chemicals are tested for long periods. When properly designed, experimental ponds should be self-sustaining. Therefore, long-term experiments can be performed, although significant decrease in diversity and productivity and increase in consanguinity undoubtedly occur in experiments that extend over several years.

Whatever the system used, duration of the experiments should be sufficient to identify both direct and indirect effects on populations and communities. Thus, studies should be performed in such a way that most plant and animal species would be able to grow and reproduce. For ponds, one year appears to be a minimum experimental duration, although Dewey and deNoyelles (1994) suggest that a duration of several years would be ideal because such long-term experiments are more likely to provide reliable information on the resistance and resilience of the exposed system and its components.

Measured Parameters: How Can Mesocosms Improve the Measurement and Significance of Classical Endpoints?

Among the various properties of natural ecosystems, stability, i.e., the ability of the ecosystem to return to an equilibrium state once it has been disturbed, seems to be an appropriate measure of ecosystem response to toxicants because it integrates both ecosystem structure and function (Johnson et al. 1994). However, this is very difficult to consider in the specific context of regulatory requirements. To be useful, this approach implies that both components of the ecosystem and functional interactions between them can be accurately described using reliable descriptors. Both ecological and biochemical or physiological parameters can be used as descriptors of contaminant effects at the ecosystem and individual levels, respectively. Ecological descriptors characterize ecosystem structure and functioning on the basis of population and community parameters and from measurements of physicochemical parameters (Larsen et al. 1986; Fairchild et al. 1992; Brock and Budde 1994; Kersting 1994). Many biochemical and physiological descriptors measured in individuals are commonly used as biomarkers because they respond to chemical exposure and may elicit further metabolic changes (McCarthy and Shugart 1990; Huggett et al. 1992; Peakall 1992; Lagadic et al. 1997, 1998). Mesocosms provide controlled conditions of exposure and facilitate access to a number of those parameters, so that causal links can be established between individual responses and changes at population and community levels that may eventually affect ecosystem structure and functioning. When investigated in natural ecosystems, such cause-and-effect relationships usually remain hypothetical because unknown factors can exert confounding influence (Hawkins et al. 1994; Maltby 1994; Cairns and Niederlehner 1995; Shaw and Kennedy 1996). In this respect, mesocosms have been proved particularly useful in experimental steps of the validation of biomarkers in fish and invertebrates (Tana et al. 1994; Bankey et al. 1995; Baturo et al. 1995; Baturo and Lagadic 1996; Vethaak et al. 1996).

IV. Future Developments
A. Predictive Role of Mesocosms

Realism, Representativity, and Replicability

Realism, representativity, and replicability of mesocosms are critical in evaluating their usefulness in both risk and impact assessment procedures. Realism is one of the main pitfalls of mesocosm design and use as it refers to the ability to reproduce natural ecosystems. Obviously, this is not a straightforward task. Beside global common characteristics that mostly depend on climatic and geological conditions at the regional scale, each natural ecosystem is unique because its structure and functioning mainly depend on local factors. Therefore, there is a conceptual opposition between realism and replicability when applied to mesocosms. Indeed, a completely realistic system, whether natural or reconstituted, is unique in space and time and thus not replicable. Considering the objectives of most studies carried out in mesocosms, replicability should be preferred to realism. Replicability may be achieved, in part, by a relative simplification of the systems. However, structural and functional characteristics of mesocosms should remain representative of those of natural ecosystems. Reconstituted systems do not need to exactly simulate natural conditions at all levels to be ecologically analogous, but key features at both structural and functional levels should be preserved as they ensure ecological representativity.

Artificial streams have relatively low ecological realism because interactions with surrounding riparian systems are not taken into account. Moreover, these systems are generally isolated from upstream systems whereas the continuum between upstream and downstream parts is a basic feature of lotic ecosystems (Vannote et al. 1980; Thorp and Delong 1994; Allan 1995). However, artificial streams provide ideal conditions to monitor the behavioral response of macroinvertebrates to contaminant exposure such as drifting (Crossland et al. 1992; Belanger 1997). Drifting is a common feature of aquatic macroinvertebrates, especially insect larvae, that is frequently enhanced by exposure to chemicals (Everts et al. 1983; Wallace et al. 1986, 1987; Krentzweiser and Kingsbury 1987; Muirhead-Thomson 1987). Quantification of this phenomenon is frequently difficult to perform in the field because of stream characteristics (depth, current velocity) and confounding factors (spates, behavioral drift in response to predators).

Reliability of information on ecotoxicological effects of chemicals tested in mesocosms closely depends on the representativity of biological processes or structures that are likely to be affected. Extrapolation from small experimental systems to the real world seems generally more problematic than with larger systems in which more complex interactions usually develop (Crossland 1994). One of the objectives of current studies performed in mesocosms is to simplify bound ecosystems in such a way as to reduce unexplained variability or to isolate mechanisms without invalidating the conclusions or predictions that can be made (Graney et al. 1995).

Modeling

Mesocosms can be very useful in the development and refinement of the various kinds of models, such as ecosystem (Brinkman et al. 1994), fate (Schramm and Goss 1990; Ratte et al. 1994), or effects models (Hanratty and Stay 1994; Hommen and Ratte 1994; Ratte et al. 1994). Among the various approaches evaluated using mesocosms, structural equation modeling seems to be promising as it may be used to identify and characterize accurate descriptors of ecosystems properties such as stability (Huggins et al. 1994; Johnson et al. 1994; Ratte et al. 1994). Mesocosms can also be used to determine the influence of variable aggregation on model outputs (Johnson et al. 1994).

B. Use of Mesocosms for Ecotoxicological Risk Assessment

Regulatory Prospects

Environmental risk characterization of chemicals in the aquatic environment is generally based on the comparison between the estimated concentrations in various compartments of ecosystems (i.e., the EECs [estimated environmental concentrations] or PECs [predicted environmental concentrations]) and the estimates of concentrations that will not cause any effect on living organisms (i.e., the NOECs [no-observed-effect concentrations]), the ratio of NOECs to EECs or PECs being defined as margins of safety (Crossland 1994; Matthiessen 1994; Van Leeuwen et al. 1994; Suter 1995). In this context, mesocosms can be used to refine estimates of NOECs, through the simultaneous exposure of many species belonging to various taxa, and EECs (Crossland et al. 1992; Fairchild et al. 1993; Taylor et al. 1994), especially through the validation of fate models (Crossland et al. 1986). In particular, mesocosms can be used for chemicals that are released in large quantities in the environment because of widespread use or which have been identified as potentially hazardous (Graney et al. 1995; Ward et al. 1995). The development of accurate fate models using such systems could significantly enhance this evaluation through refining the estimation of EECs (Crossland 1994). In addition, decision-making tools are currently being developed based upon long-term comprehensive investigations of the effects of selected chemicals on various physicochemical and biological parameters in mesocosms (Belanger et al. 1999).

During the last half of the 1980s, guidelines were published in the U.S. that meet the requirements of both the Federal Insecticide, Fungicide, and Rodenticide Act (FIFRA) and the Toxic Substances Control Act (TSCA) (Touart 1988). In the context of FIFRA, specific guidance for field testing of pesticides was provided for wildlife (Fite et al. 1988) and aquatic organisms (Touart 1988). At the beginning of the 1990s, the Office of Pesticide Programs (OPP) stimulated reflection on this approach (SETAC-RESOLVE 1992; RESOLVE 1992), and tests using large-scale mesocosms were abandoned in 1992 mainly because their cost-effectiveness was questionable (Graney et al. 1994; Boyle and Fairchild 1997). According to the current regulation, potential adverse ecological effects of pesticides are collected following a tiered approach that begins with low-cost laboratory toxicity tests and continues, if necessary, with more expensive chronic toxicity tests. Simulated or actual field studies, e.g., mesocosm studies, are the final tier of this approach (Nabholz et al. 1997; Touart and Maciorowski 1997). Such studies may be performed to experimentally confirm the safety of a chemical under anticipated conditions of use or if the margin of safety is low. However, such studies are not compulsory because "the OPP intends to regulate on laboratory and exposure information without requiring or waiting for field studies except in rare circumstances" (Touart and Maciorowski 1997). The USEPA Office of Prevention, Pesticides and Toxic Substances (OPPTS) has developed a new series of ecological effects test guidelines that includes a proposal for Field Testing for Aquatic Organisms (OPPTS 850.1950; USEPA-OPPTS 1996) based on the use of pond mesocosms, with a minimum of four experimental treatments with at least three replicates per treatment level. Additional details such as size, depth, and technical features are also provided in this proposal.

In the European Union, environmental risk characterization of pesticides follows the 91/414/EEC directive, which is also based on a tiered approach. Mesocosm studies may be required when compounds do not succeed in passing the first tier of aquatic risk assessment. Regulatory procedures for higher-tier risk assessment are still under development. In this context, harmonization of the approaches used to characterize risks in the EU (European Union) and the U.S. is promoted by the OECD. Expert groups are currently working on guidance documents for such higher-tier aquatic risk characterization (Campbell et al. 1999). Moreover, a new OECD guideline for freshwater lentic field tests similar to the USEPA-OPPTS document is under preparation, based on the draft version (OECD 1996), which was open for comment on the Organization web site.

Mesocosms in Alternative Testing Strategies of Environmental Chemicals
The SGOMSEC 13 meeting held at the EC Joint Research Center in Ispra, Italy, recently addressed current practices and future developments in alternative testing of chemicals (Stokes et al. 1998). The Ecotoxicology Group initiated discussion on how ecotoxicity testing in mesocosms meets the essential concern of alternative methodologies (Walker et al. 1998):

Refinement. Certainly, the use of mesocosms refines the classical methods of ecotoxicological risk assessment because they provide conditions for a better understanding of environmentally relevant effects of chemicals. Indeed, mesocosms provide a more realistic approach for the evaluation of effects of chemicals at many different levels of organization, from the molecule to population and community, and for different types of organisms, from bacteria to invertebrates and lower vertebrates. Mesocosms offer the ability to assess effects of contaminants by looking at the parts (individuals, populations, communities) and the whole (communities, ecosystems) simultaneously. They also appear as potent tools to predict changes at the highest levels of organization (population, community, and ecosystem) from measurements of individual endpoints.

Replacement. Ecotoxicological investigations in mesocosms of course will never entirely replace the use of laboratory animals. However, mesocosms allow the tests to be performed on species that are not of major societal concern but which play key roles in the structure and functioning of ecosystems. For example, investigations of chemical effects in freshwater mesocosms have largely used invertebrate species because of their importance in aquatic food webs (Lagadic and Caquet 1998). For species replacement, there is an important need for interspecies comparisons of toxicological effects at different levels (molecular, cellular) within the individual. A given chemical may elicit different individual expression of effects in vertebrates and lower organisms (invertebrates, microorganisms); the establishment of corresponding endpoints between species would help the identification of cross-references.

Reduction. To some extent, investigations of ecotoxicological effects in mesocosms can significantly reduce the need for animals when, in a particular test, ecosystem-relevant functional endpoints can be measured. Among those endpoints, plankton respiration, phytoplankton photosynthesis, concentrations of chlorophyll a, pH, dissolved oxygen or nitrogen, and ammonia are the most commonly measured in aquatic ecosystems.

The need for using animals in ecotoxicity testing in mesocosms clearly depends on the endpoints that must be assessed. In this respect, mesocosms allow nondestructive measurements of integrated endpoints at high levels of organization or functional endpoints. Chemical tests in mesocosms should therefore be designed to reduce, or in some cases even replace, the use of vertebrates, or to reduce the amount of suffering of vertebrates by measuring nondestructive parameters (Walker 1998).

V. Conclusions

Aquatic mesocosms have been used in ecotoxicology for approximately 20 yr and were sometimes claimed to be essential tools, especially for regulatory purposes. However, much criticism rapidly arose concerning their real scientific value mainly because of high experiment cost and poor repeatability and reproducibility. In fact, it is now possible to identify some key features of mesocosm use that greatly improve the value of such systems in ecotoxicology.

Mesocosms constitute valuable tools in both experimental and applied ecotoxicology. From a fundamental point of view, aquatic mesocosms can be used to identify and quantify links between changes measured at the individual level and their consequences at population and community levels. They also give the opportunity to evaluate the effects of environmental contaminants on ecological processes that cannot be assessed in laboratory experiments, e.g., indirect effects. In the field of applied ecotoxicology, e.g., regulatory concern, they may be used to test laboratory results or field observations, especially for persistent chemicals.

Because the cost of experiments and the difficulty of data interpretation increase with the size of the systems, it has become evident that mesocosm design should promote systems of a "reasonable" size. This criterion refers to the ability of the systems to ensure that sampling will not result in a more important impact than contamination and that the systems will be self-sustainable, i.e., that they will ensure the development of every compartment without human intervention during the experiment. This objective can be attained using a design that promotes structuring processes identified in ecological studies and which continuously refers to ecological theory. Mesocosm communities should not merely be the result of the juxtaposition of species chosen only because they constitute valuable models or because they are easily reared in the laboratory.

Because each natural ecosystem is a unique result of the combination and interaction of many biotic and abiotic factors, it makes no sense to define a standardized mesocosm. However, standard protocols, or at least guidelines, are required to ensure that critical parameters or processes are not neglected. Moreover, because all the possible endpoints cannot reasonably be assessed, emphasis should be placed on rationalizing data collection and processing. In this context, modeling could significantly improve mesocosm experiments through identification of integrative parameters.

Summary

Mesocosms have been used in aquatic ecotoxicology for approximately 20 years and were sometimes claimed to be essential tools, especially for regulatory purposes. The term aquatic mesocosm currently describes indoor and outdoor artificial streams or experimental ponds and enclosures. The use of mesocosms refines the classical methods of ecotoxicological risk assessment because mesocosms provide conditions for a better understanding of environmentally relevant effects of chemicals. They make it possible to assess effects of contaminants by looking at the parts (individuals, populations, communities) and the whole (ecosystems) simultaneously. Ecotoxicological investigations in mesocosms will not entirely replace the use of laboratory animals. However, they allow tests to be performed on species that are not of major societal concern, but which play key roles in the structure and function of ecosystems. In this respect, mesocosms allow nondestructive measurements of integrated endpoints. They also appear as potent tools to predict changes at the highest levels of organization (population, community, and ecosystem) from measurements of individual endpoints. However, after a period of extensive use, regulatory studies using large-scale mesocosms were more or less abandoned at the beginning of the 1990s, mainly because their cost-effectiveness was questionable. This review covers key features of outdoor aquatic mesocosms that can be critical for their use in environmental risk assessment of chemicals and emphasizes the optimization of their use for such purpose.

The originality of mesocosms is mainly based on the combination of ecological realism, achieved by introduction of the basic components of natural ecosystems, and facilitated access to a number of physicochemical, biological, and toxicological parameters that can be controlled to some extent. This characteristic determines various features of the systems such as the minimal size required, initial physicochemical and biological composition, or choice of model species for ecotoxicological investigations.

Ecological maturity of mesocosms affects the degree of variability of both physicochemical and biological parameters used to investigate the impact of contaminants. Adequate time is required to establish a number of interacting functional groups. The choice of appropriate time scales must be considered in the selection of both study duration and sampling frequency. Whatever the system used, duration of experiments should be sufficient to identify both direct and indirect effects on populations and communities. The choice of the experimental design should be based on the objectives of the study rather than on theoretical considerations. In addition to classical parametric statistical methods, nonparametric approaches and multivariate analysis may significantly improve data processing.

Realism, representativity, and replicability of mesocosms are critical for evaluating their usefulness in both risk and impact assessment procedures. Each natural ecosystem is unique because its structure and function mainly depend on local factors. Therefore, there is a conceptual opposition between realism and replicability when applied to mesocosms. Considering the objectives of most mesocosm studies, replicability should be preferred to realism. Replicability may be achieved, in part, by a relative simplification of the systems. Reconstituted systems do not need to exactly simulate natural conditions at all levels, but key features at both structural and functional levels should be preserved as they ensure ecological representativity.

Reliability of information on ecotoxicological effects of chemicals tested in aquatic mesocosms closely depends on the representativity of biological processes or structures that are likely to be affected. Extrapolation from small experimental systems to the real world seems generally more problematic than the use of larger systems in which more complex interactions usually occur. One of the objectives of current practices performed in mesocosms is to make simplifications of bound ecosystems to reduce unexplained variability or to isolate mechanisms without invalidating the conclusions or predictions that can be made. Mesocosms can thus be very useful in the development and refinement of various kinds of models such as ecosystem, fate, or effects models. In the context of risk assessment of chemicals, aquatic mesocosms can be used to refine estimates of the expected concentrations in various compartments of aquatic ecosystems and of concentrations that will not cause any effect on living organisms.

Because each natural ecosystem is a unique result of the combination and interactions of many biotic and abiotic factors, it is not practical to define a standardized mesocosm. However, standard protocols, or at least guidelines, are required to ensure that critical parameters or processes are not neglected. Moreover, as all the possible endpoints cannot reasonably be assessed, emphasis should be on rationalizing data collection and processing. In this context, modeling could significantly improve mesocosm experiments through the identification of integrative parameters.

Acknowledgments

The authors are grateful to SGOMSEC 13 Meeting (Alternative Testing Methodologies) contributors for helpful and stimulating discussions. They also thank Dr. Tom La Point and Prof. John Cairns, Jr., who provided valuable comments on the manuscript.

References

Adams JN, Beschta RL (1980) Gravel bed composition in Oregon coastal streams. Can J Fish Aquat Sci 37:1514–1521.

Ali A, Stanley BH (1982) Effects of a new carbamate insecticide larvin (UC-51762) on some nontarget aquatic invertebrates. Fla Entomol 65:477–483.

Allan JD (1995) Stream Ecology. Structure and Function of Running Waters. Chapman & Hall, London.

Allen KN (1991) Seasonal variation of selenium in outdoor experimental stream-wetland systems. J Environ Qual 20:865–868.

Ammann LP, Waller WT, Kennedy JH, Dickson KL, Mayer FL (1997) Power, sample size and taxonomic sufficiency for measures of impact in aquatic systems. Environ Toxicol Chem 16:2421–2431.

Andersson G, Berggren H, Cronberg G, Gelin C (1978) Effects of planktivorous and benthivorous fish on organisms and water chemistry in eutrophic lakes. Hydrobiologia 59:9–15.

Angerilli NPD, Beirne BP (1980) Influences of aquatic plants on colonization of artificial ponds by mosquitoes and their insect predators. Can Entomol 112:793–796.

Arnold JG, Williams JR, Nicks AD, Sammons NB (1990) SWRRB: a basin scale simulation model for soil and water resources management. Texas A & M University Press, College Station, TX.

Arthur JW, (1988) Application of laboratory-devised criteria to an outdoor stream ecosystem. Int J Environ Stud 32:97–110.

Arthur JW, Zischke JA, Allen KN, Hermanutz RO (1983) Effects of diazinon on macroinvertebrates and insect emergence in outdoor experimental channels. Aquat Toxicol 4:283–301.

Arumugam PT, Geddes MC (1986) An enclosure for experimental field studies with fish and zooplankton communities. Hydrobiologia 135:215–221.

Bakke T, Follum OA, Moe KA, Sorensen K (1988) The GEEP workshop: mesocosm exposures. Mar Ecol Prog Ser 46:13–18.

Bankey LA, Van Veld PA, Borton DL, LaFleur L, Stegeman JJ (1995) Responses of cytochrome P4501A in freshwater fish exposed to bleached kraft mill effluent in experimental stream channels. Can J Fish Aquat Sci 52:439–447.

Barnes LE (1983) The colonization of ball-clay ponds by macroinvertebrates and macrophytes. Freshwater Biol 132:561–578.

Baturo W, Lagadic L (1996) Benzo[a]pyrene hydroxylase and glutathione S-transferase activities as biomarkers in *Lymnaea palustris* (Mollusca, Gastropoda) exposed to atrazine and hexachlorobenzene in freshwater mesocosms. Environ Toxicol Chem 15: 771–781.

Baturo W, Lagadic L, Caquet T (1995) Growth, fecundity, and glycogen utilization in *Lymnaea palustris* exposed to atrazine and hexachlorobenzene in freshwater mesocosms. Environ Toxicol Chem 14:503–511.

Belanger SE (1994) Review of experimental microcosm, mesocosm, and field tests used to evaluate the potential hazard of surfactants to aquatic life and the relation to single species data. In: Hill IR, Heimbach F, Leeuwangh P, Matthiessen P (eds) Freshwater Field Tests for Hazard Assessment of Chemicals. Lewis, Boca Raton, pp 287–314.

Belanger SE (1997) Literature review and analysis of biological complexity in model stream ecosystems: influence of size and experimental design. Ecotoxicol Environ Saf 36:1–16.

Belanger SE, D'Angelo DJ, Bowling JW (1999) Decision making tools useful in developing conclusions from model ecosystem tests. Abstract n°2t/P014, 9th Annual Meeting of SETAC-Europe, 25–29 May 1999, Leipzig.

Beyers RJ, Odum HT (1993) Ecological Microcosms. Springer-Verlag, New York.

Blake G (1994) Are aquatic macrophytes useful in field tests? In: Hill IR, Heimbach F, Leeuwangh P, Matthiessen P (eds) Freshwater Field Tests for Hazard Assessment of Chemicals. Lewis, Boca Raton, pp 183–189.

Bowling JW, Giesy JP, Kania HJ, Knight RL (1980) Large-scale microcosms for assessing fates and effects of trace contaminants. In: Giesy JP Jr (ed) Microcosms in Ecological Research. U.S. Dept. of Energy Symposium Series 52. National Technical Information Workshop, Blacksburg, VA, pp 224–247.

Boyle TP, Fairchild JF (1997) The role of mesocosm studies in ecological risk analysis. Ecol Appl 7:1099–1102.

Boyle TP, Fairchild JF, Robinson-Wilson EF, Haverland PS, Lebo JA (1996) Ecological restructuring in experimental aquatic mesocosms due to the application of diflubenzuron. Environ Toxicol Chem 15:1806–1814.

Brabrand Å, Faafeng B, Nilssen JPM (1987) Pelagic predators and interfering algae: stabilizing factors in temperate eutrophic lakes. Arch Hydrobiol 110:533–552.

Brazner JC, Kline ER (1990) Effects of chlorpyrifos on the diet and growth of larval fathead minnows, *Pimephales promelas*, in littoral enclosures. Can J Fish Aquat Sci 47:1157–1165.

Brazner JC, Heinis LJ, Jensen DA (1989) A littoral enclosure for replicated field experiments. Environ Toxicol Chem 8:1209–1216.

Brinkman AG, Philippart CJM, Holtrop G (1994) Mesocosms and ecosystem modelling. Vie Milieu 44:29–37.

Brock TCM, Budde BJ (1994) On the choice of structural parameters and endpoints to indicate responses of freshwater ecosystems to pesticide stress. In: Hill IR, Heimbach F, Leeuwangh P, Matthiessen P (eds) Freshwater Field Tests for Hazard Assessment of Chemicals. Lewis, Boca Raton, pp 19–56.

Brooks JL, Dodson SI (1965) Predation, body size, and composition of plankton. Science 150:28–35.

Burns LA (1990) Exposure Analysis Modeling System: User's Guide for EXAMS II Version 2.94. EPA/600/3-89-084. U.S. E.P.A. Office of Research and Development, Washington, DC.

Burns LA, Cline DM (1985) Exposure Analysis Modeling System: Reference Manual for EXAMS II. EPA-600/3-85-038. U.S. E.P.A. Office of Research and Development, Washington, DC.

Burns LA, Cline DM, Lassiter RR (1982) Exposure Analysis Modeling System (EXAMS): User Manual and System Documentation. EPA-600/3-82-023. U.S. E.P.A. Office of Research and Development, Washington, DC.

Burton GA (1996) Artificial sediments: current issues for standardization. OECD Expert Group on Chironomidae Testing, OECD, Paris.

Cairns J Jr (1988) Putting the eco in ecotoxicology. Reg Toxicol Pharmacol 8:226–238.

Cairns J Jr, Niederlehner BR (1995) Ecological Toxicity Testing. Scale, Complexity, and Relevance. Lewis, Boca Raton.

Cairns J Jr, Bidwell JR, Arnegard ME (1996) Toxicity testing with communities: microcosms, mesocosms and whole-system manipulations. Rev Environ Contam Toxicol 147:45–69.

Campbell PJ, Arnold DJS, Brock TCM, Grandy NJ, Heger W, Heimbach F, Maund SJ,

Streloke M (1999) Guidance Document on Higher-tier Aquatic Risk Assessment for Pesticides. SETAC-Europe, Bussels.

Caquet T, Thybaud E, Le Bras S, Jonot O, Ramade F (1992) Fate and biological effects of lindane and deltamethrin in freshwater mesocosms. Aquat Toxicol 23:261–278.

Caquet T, Lagadic L, Jonot O, Baturo W, Kilanda M, Simon P, Le Bras S, Echaubard M, Ramade F (1996) Outdoor experimental ponds (mesocosms) designed for long-term ecotoxicological studies in aquatic environment. Ecotoxicol Environ Saf 34: 125–133.

Carpenter SR, Kitchell JF, Hodgson JR (1985) Cascading trophic interactions and lake productivity. BioScience 35:634–639.

Carsel RF, Smith CN, Mulkey LA, Dean JD, Jowise P (1984) Users Manual for the Pesticide Root Zone Model (PRZM). EPA 600/3-84-109. U.S. E.P.A., Washington, DC.

Chant L, Cornett RJ (1988) Measuring contaminant transport rates betwen water and sediments using limnocorrals. Hydrobiologia 159:237–245.

Christman VD, Voshell JR Jr (1993) Changes in the benthic macroinvertebrate community in two years of colonization of new experimental ponds. Int Rev Gesamten Hydrobiol 78:481–491.

Christman VD, Voshell JR Jr, Jenkins DG, Rosenzweig MS, Layton RJ, Buikema AL Jr (1994) Ecological development and biometry of untreated pond mesocosms. In: Graney RL, Kennedy JH, Rodgers JH Jr (eds) Aquatic Mesocosm Studies in Ecological Risk Assessment. Lewis, Boca Raton, pp 105–127.

Cid Montañás JF, Van Hattum B, Deneer J (1995) Bioconcentration of chlorpyrifos by the freshwater isopod *Asellus aquaticus* (L.) in outdoor experimental ditches. Environ Pollut 88:137–146.

Clark JR, Noles JL (1994) Contaminant effects in marine/estuarine systems: field studies and scaled simulations. In: Graney RL, Kennedy JH, Rodgers JH Jr (eds) Aquatic Mesocosm Studies in Ecological Risk Assessment. Lewis, Boca Raton, pp 47–68.

Clarke KR (1999) Nonmetric multivariate analysis in community-level ecotoxicology. Environ Toxicol Chem 18:118–127.

Clements WH, Cherry DS, Cairns J Jr (1988) Impact of heavy metals on insect communities in streams: a comparison of observational and experimental results. Can J Fish Aquat Sci 45:2017–2025.

Clements WH, Farris JL, Cherry DS, Cairns J Jr (1989) The influence of water quality on macroinvertebrate community responses to copper in outdoor experimental streams. Aquat Toxicol 14:249–262.

Clements WH, Cherry DS, Cairns J Jr (1990) Macroinvertebrate community response to copper in laboratory and field experimental streams. Arch Environ Contam Toxicol 19:361–365.

Crossland NO (1982) Aquatic toxicology of cypermethrin. II. Fate and biological effects in pond experiments. Aquat Toxicol 2:205–222.

Crossland NO (1984) Fate and biological effects of methyl parathion in outdoor ponds and laboratory aquaria. II: Effects. Ecotoxicol Environ Saf 8:482–495.

Crossland NO (1990) The role of mesocosm studies in pesticide registration. In: Proceedings, Brighton Crop Protection Conference, Brighton, UK, pp 499–508.

Crossland NO (1994) Extrapolating from mesocosms to the real world. Toxicol Ecotoxicol News 1:15–22.

Crossland NO, Bennett D (1984) Fate and biological effects of methyl parathion in outdoor ponds and laboratory aquaria. I. Fate. Ecotoxicol Environ Saf 8:471–481.

Crossland NO, Wolff CJ (1985) Fate and effects of pentachlorophenol in outdoor ponds. Environ Toxicol Chem 4:73–86.

Crossland NO, Bennett D (1989) Outdoor ponds: their use to evaluate the hazards of organic chemicals in aquatic environments. In: Boudou A, Ribeyre F. (eds) Aquatic Ecotoxicology: Fundamental Concepts and Methodologies, Vol. I. CRC Press, Boca Raton, pp 273–296.

Crossland NO, Bennett D, Wolff CJM, Swannell RPJ (1986) Evaluation of models to assess the fate of chemicals in aquatic systems. Pestic Sci 17:297–304.

Crossland NO, Mitchell GC, Bennett D, Maxted J (1991) An outdoor artificial stream system designed for ecotoxicological studies. Ecotoxicol Environ Saf 22:175–183.

Crossland NO, Mitchell GC, Dorn PB (1992) Use of outdoor artificial streams to determine threshold toxicity concentrations for a petrochemical effluent. Environ Toxicol Chem 11:49–59.

Crum SJH, Brock TCM (1994) Fate of chlorpyrifos in indoor microcosms and outdoor experimental ditches. In: Hill IR, Heimbach F, Leeuwangh P, Matthiessen P (eds) Freshwater Field Tests for Hazard Assessment of Chemicals. Lewis, Boca Raton, pp 315–322.

Cuffney TM, Hart DC, Wolbach KC, Wallace JB, Lugthart GJ, Smith-Cuffney FL (1990) Assessment of community and ecosystem level effects in lotic environments: the role of mesocosms and field studies. In: Cuffney TF (ed) Experimental Ecosystems: Applications to Ecotoxicology. Technical Information Workshop, North American Benthological Society, Virginia Polytechnic Institute and State University, Blacksburg, VA.

Cushman RM, Goyert JC (1984) Effects of a synthetic crude oil on pond benthic insects. Environ Pollut Ser A 33:163–186.

Day KE, Kaushik NK, Solomon KR (1987) Impact of fenvalerate on enclosed freshwater planktonic communities and on *in situ* rates of filtration of zooplankton. Can J Fish Aquat Sci 44:1714–1728.

deNoyelles F, Kettle WD, Sinn DE (1982) The responses of plankton communities in experimental ponds to atrazine, the most heavily used pesticide in the United States. Ecology 63:1285–1293.

deNoyelles F Jr, Kettle WD, Fromm CH, Moffett MF, Dewey SL (1989) Use of experimental ponds to assess the effects of a pesticide on the aquatic environment. In: Voshell JR Jr (ed) Using Mesocosms to Assess the Aquatic Ecological Risk of Pesticides. Misc Publ Entomol Soc Am 75:41–56.

Detenbeck NE, Hermanutz R, Allen K, Swift MC (1996) Fate and effects of the herbicide atrazine in flow-through wetland mesocosms. Environ Toxicol Chem 15:937–946.

Deutsch WG, Webber EC, Bayne DR, Reed CW (1992) Effects of largemouth bass stocking rate on fish populations in aquatic mesocosms used for pesticide research. Environ Toxicol Chem 11:5–10.

Dewey SL (1986) Effects of the herbicide atrazine on aquatic insect community structure and emergence. Ecology 67:148–162.

Dewey SL, deNoyelles F Jr (1994) On the use of ecosystem stability measurements in ecological effects testing. In: Graney RL, Kennedy JH, Rodgers JH Jr (eds) Aquatic Mesocosm Studies in Ecological Risk Assessment. Lewis, Boca Raton, pp 605–625.

Dorn P, Rodgers JH Jr, Dubey ST, Gillespie WB Jr, Lizotte RE Jr (1996) An assessment of the ecological effects of C_{9-11} linear ethoxylate surfactant in stream mesocosm experiments. Ecotoxicology 6:275–282.

Dorn P, Rodgers JH Jr, Gillespie WB Jr, Lizotte RE Jr, Dunn AW (1997) The effects of

na C_{12-13} linear alcohol ethoxylate surfactant on periphyton, macrophytes, invertebrates and fish in stream mesocosms. Environ Toxicol Chem 16:1634–1645.
Drenner RW, Threlkeld ST, McCraken MD (1986) Experimental analysis of the direct and indirect effects of an omnivorous filter-feeding clupeid on plankton community structure. Can J Fish Aquat Sci 43:1935–1945.
Drent J, Kersting K (1993) Experimental ditches for ecotoxicological experiments and eutrophication research under natural conditions. Water Res 27:1497–1500.
Everts, JM, van Frankenhuyzen K, Roman B, Koeman JH (1983) Side-effects of experimental pyrethroid applications for the control of tsetse flies in a riverine forest habitat (Africa). Arch Environ Contam Toxicol 12:91–97.
Fairchild JF, Little EE (1993) Use of mesocosm studies to examine direct and indirect impacts of water quality on early life stages of fishes. Am Fish Soc Symp 14:95–103.
Fairchild JF, Boyle T, English RW, Rabeni C (1987) Effects of sediment and contaminated sediment on structural and functional components of experimental stream ecosystems. Water Air Soil Pollut 36:271–293.
Fairchild JF, La Point TW, Zajicek JL, Nelson MK, Dwyer FJ, Lovely PA (1992) Population-, community-, and ecosystem-level responses of aquatic mesocosms to pulsed doses of a pyrethroid insecticide. Environ Toxicol Chem 11:115–129.
Fairchild JF, Dwyer FJ, La Point TW, Burch SA, Ingersoll CW (1993) Evaluation of a laboratory-generated NOEC for linear alkylbenzene sulfonate in outdoor experimental streams. Environ Toxicol Chem 10:1763–1776.
Fairchild JF, La Point TW, Schwartz TR (1994) Effects of an herbicide and insecticide mixture in aquatic mesocosms. Arch Environ Contam Toxicol 27:527–533.
Farke H, Wonneberger K, Gunkel W, Dahlmann G (1985) Effects of oil and a dispersant on intertidal organisms in field experiments with a mesocosm, the Bremerhaven Caisson. Mar Environ Res 15:97–114.
Farmer D, Hill IR, Maund SJ (1995) A comparison of the fate and effects of two pyrethroid insecticides (lambda-cyhalothrin and cypermethrin) in pond mesocosms. Ecotoxicology 4:219–244.
Ferrington LC, Blackwood MA, Wright CA, Anderson TM, Goldhammer DS (1994) Sediment transfers and representativeness of mesocosm tests fauna. In: Graney RL, Kennedy JH, Rodgers JH Jr (eds) Aquatic Mesocosm Studies in Ecological Risk Assessment. Lewis, Boca Raton, pp 179–200.
Fite EC, Turner LW, Cook NJ, Stunkard C (1988) Guidance Document for Conducting Terrestrial Field Studies. US-EPA/540/09-88-109. U.S. E.P.A., Washington, DC.
Francis DR, Kane TC (1995) Effect of substrate on colonization of experimental ponds by Chironomidae (Diptera). J Freshwater Ecol 10:57–63.
Ganzelmeier H. (1997) Abtrift und Bodenblastang beim Ausbringen von Pflanzenschatzmitteln. Mitt Biol Bundes Land Forst Berlin 328:115–124.
Gelwick FP, Matthews WJ (1993) Artificial streams for studies of fish ecology. J North Am Benthol Soc 12:343–347.
Genter RB, Cherry DS, Smith EP, Cairns J Jr (1987) Algal-periphyton population and community changes from zinc stress in stream mesocosms. Hydrobiologia 153:261–275.
Getsinger KD, Dick GO, Crouch RM, Nelson LS (1994) Mesocosm evaluation of bensulfuron methyl activity on Eurasian watermilfoil, *Vallisneria*, and American pondweed. J Aquat Plant Manage 32:1–6.
Giddings JM, Franco PJ, Cushman RM, Hook LA, Southworth GR, Stewart AJ (1984) Effects of chronic exposure to coal-derived oil on freshwater ecosystems. II. Experimental ponds. Environ Toxicol Chem 3:465–488.

Giddings JM, Biever RC, Annunziato MF, Hosmer AJ (1996) Effects of diazinon on large outdoor pond microcosms. Environ Toxicol Chem 15:618–629.

Giddings JM, Biever RC, Racke KD (1997) Fate of chlorpyrifos in outdoor pond microcosms and effect on growth and survival of bluegill sunfish. Environ Toxicol Chem 16:2353–2362.

Giesy JP Jr, Odum EP (1980) Microcosmology: introductory comments. In: Giesy JP Jr (ed) Microcosms in Ecological Research. U.S. Dept. of Energy Symposium Series 52. National Technical Information Workshop, Blacksburg, VA, pp 1–13.

Gilinsky E (1984) The role of fish predation and spatial heterogeneity in determining benthic community structure. Ecology 65:455–468.

Gillespie WB Jr, Rodgers JH Jr, Crossland NO (1996) Effects of a nonionic surfactant (C_{14-15} AE-7) on aquatic invertebrates in outdoor stream mesocosms. Environ Toxicol Chem 15:1418–1422.

Gillespie WB Jr, Rodgers JH Jr, Dorn PB (1997) Responses of aquatic invertebrates to a C_{9-11} non-ionic surfactant in outdoor stream mesocosms. Aquat Toxicol 37:221–236.

Graney RL, Giesy JP Jr, DiToro D (1989) Mesocosm experimental design strategies: advantages and disadvantages in ecological risk assessment. In: Voshell JR Jr (ed) Using Mesocosms to Assess the Aquatic Ecological Risk of Pesticides. Misc Publ Entomol Soc Am 75:74–88.

Graney RL, Kennedy JH, Rodgers JH Jr (1994) Aquatic Mesocosm Studies in Ecological Risk Assessment. Lewis, Boca Raton.

Graney RL, Giesy JP Jr, Clark JR (1995) Field studies. In: Rand GM (ed) Fundamentals of Aquatic Toxicology, 2nd Ed. Taylor & Francis, Washington, DC, pp 257–305.

Groenendijk P, van der Kolk JWH, Travis KZ (1994) Prediction of exposure concentrations in surface waters. In: Hill IR, Heimbach F, Leeuwangh P, Matthiessen P (eds) Freshwater Field Tests for Hazard Assessment of Chemicals. Lewis, Boca Raton, pp 105–125.

Guckert JB (1993) Artificial streams in ecotoxicology. J North Am Benthol Soc 12:350–356.

Hall DJ, Cooper WE, Werner EE (1970) An experimental approach to the production dynamics and structure of freshwater animal communities. Limnol Oceanogr 15:859–928.

Hamer MJ, Hill IR, Rondon L, Caguan A (1994) The effects of lambda-cyhalothrin in aquatic field studies. In: Hill IR, Heimbach F, Leeuwangh P, Matthiessen P (eds) Freshwater Field Tests for Hazard Assessment of Chemicals. Lewis, Boca Raton, pp 331–337.

Hamilton PB, Lean DRS, Jackson GS, Kaushik NK, Solomon KR (1989) The effect of two applications of atrazine on the water quality of freshwater enclosures. Environ Pollut 60:291–304.

Hanazato T, Yasuno M (1987) Effects of a carbamate insecticide, carbaryl, on the summer phyto- and zooplankton communities in ponds. Environ Pollut 48:145–159.

Hanazato T, Yasuno M (1990a) Influence of *Chaoborus* density on the effects of an insecticide on zooplankton communities in ponds. Hydrobiologia 194:183–197.

Hanazato T, Yasuno M (1990b) Influence of time of application of an insecticide on recovery patterns of a zooplankton community in experimental ponds. Arch Environ Contam Toxicol 19:77–83.

Hanazato T, Iwakuma T, Yasuno M, Sakamoto M (1989) Effects of temephos on zooplankton communities in enclosures in a shallow eutrophic lake. Environ Pollut 59:305–314.

Hanratty MP, Stay FS (1994) Field evaluation of the Littoral Ecosystem Risk Assessment Model's predictions of the effects of chlorpyrifos. J Appl Ecol 31:439–453.

Harrelson RA, Rodgers JH Jr, Lizotte RE Jr, Dorn PB (1997) Responses of fish exposed to a C_{9-11} linear alcohol ethoxylate nonionic surfactant in stream mesocosms. Ecotoxicology 6:321–333.

Hatakeyama S, Shiraishi H, Uno S (1997) Overall pesticide effects on growth and emergence of two species of Ephemeroptera in a model stream carrying pesticide-polluted river water. Ecotoxicology 6:167–180.

Havens KE (1995) Insecticide (carbaryl, 1-naphthyl-N-methylcarbamate) effects on a freshwater planktonic community: zooplankton size, biomass, and algal abundance. Water Air Soil Pollut 84:1–10.

Havens KE, Heath RT (1989) Acid and aluminium effects on freshwater zooplankton: an *in situ* mesocosm study. Environ Pollut 62:195–211.

Hawkins SJ, Proud SV, Spence SK, Southward AJ (1994) From the individual to the community and beyond: water quality, stress indicators and key species in coastal systems. In: Sutcliffe DW (ed) Water Quality and Stress Indicators in Marine and Freshwater Ecosystems: Linking Levels of Organisation (Individuals, Populations, Communities). Freshwater Biological Association, Ambleside, pp 35–62.

Heimbach F (1994) Methodologies of aquatic field tests: system design for field tests in still waters. In: Hill IR, Heimbach F, Leeuwangh P, Matthiessen P (eds) Freshwater Field Tests for Hazard Assessment of Chemicals. Lewis, Boca Raton, pp 141–150.

Heimbach F, Pflueger W, Ratte H-T (1992) Use of small artificial ponds for assessment of hazards to aquatic ecosystems. Environ Toxicol Chem 11:27–34.

Heimbach F, Berndt J, Pflueger W (1994) Fate and biological effects of an herbicide on two artificial pond ecosystems of different size. In: Graney RL, Kennedy JH, Rodgers JH Jr (eds) Aquatic Mesocosm Studies in Ecological Risk Assessment. Lewis, Boca Raton, pp 303–320.

Heinis LJ, Knuth ML (1992) The mixing, distribution and persistence of esfenvalerate within littoral enclosures. Environ Toxicol Chem 11:11–25.

Herman D, Kaushik NK, Solomon KR (1986) Impact of atrazine on periphyton in freshwater enclosures and some ecological consequences. Can J Fish Aquat Sci 43:1917–1925.

Hermanutz RO, Allen KN, Roush TH, Hedtke SF (1992) Effects of elevated selenium concentrations on bluegills (*Lepomis macrochirus*) in outdoor experimental streams. Environ Toxicol Chem 11:217–224.

Hill IR, Heimbach F, Leeuwangh P, Matthiessen P (1994a) Freshwater Field Tests for Hazard Assessment of Chemicals. Lewis, Boca Raton.

Hill IR, Runnalls JK, Kennedy JH, Ekoniak P (1994b) Effects of lambda-cyhalothrin on aquatic organisms in large-scale mesocosms. In: Hill IR, Heimbach F, Leeuwangh P, Matthiessen P (eds) Freshwater Field Tests for Hazard Assessment of Chemicals. Lewis, Boca Raton, pp 345–360.

Hoagland KD, Drenner RW, Smith JD, Cross DR (1993) Freshwater community responses to mixtures of agricultural pesticides: effects of atrazine and bifenthrin. Environ Toxicol Chem 12:627–637.

Hommen U, Ratte H-T (1994) Application of a plankton simulation model to outdoor-microcosm case studies. In: Hill IR, Heimbach F, Leeuwangh P, Matthiessen P (eds) Freshwater Field Tests for Hazard Assessment of Chemicals. Lewis, Boca Raton, pp 493–502.

Howick GL, Giddings JM, deNoyelles F, Ferrington LC Jr, Kettle WD, Baker D (1992)

Rapid establishment of test conditions and trophic-level interactions in 0.04-hectare earthen pond mesocosms. Environ Toxicol Chem 11:107-114.

Howick GL, deNoyelles F, Dewey SL, Mason L, Baker D (1993) The feasibility of stocking largemouth bass in 0.04-ha mesocosms used for pesticide research. Environ Toxicol Chem 12:1883-1893.

Howick GL, deNoyelles F Jr, Giddings JM, Graney RL (1994) Earthen ponds vs. fiberglass tanks as venues for assessing the impact of pesticides on aquatic environments: a parallel study with sulprofos. In: Graney RL, Kennedy JH, Rodgers JH Jr (eds) Aquatic Mesocosm Studies in Ecological Risk Assessment. Lewis, Boca Raton, pp 321-336.

Huggett RJ, Kimerle RA, Mehrle PM Jr, Bergman HL (1992) Biomarkers: Biochemical, Physiological, and Histological Markers of Anthropogenic Stress. Lewis, Boca Raton.

Huggins DG, Johnson ML, deNoyelles F Jr (1994) The ecotoxic effects of atrazine on aquatic ecosystems: an assessment of direct and indirect effects using structural equation modeling. In: Graney RL, Kennedy JH, Rodgers JH Jr (eds) Aquatic Mesocosm Studies in Ecological Risk Assessment. Lewis, Boca Raton, pp 653-692.

Hurlbert SH, Mulla MS, Keith JO, Westlake WE, Düsch ME (1970) Biological effects and persistence of Dursban® in freshwater ponds. J Econ Entomol 63:43-52.

Hurlbert SH, Mulla MS, Wilson HR (1972) Effects of an organophosphorus insecticide on the phytoplankton, zooplankton, and insect populations of fresh-water ponds. Ecol Monogr 42:269-299.

Jak RG, Mas JL, Scholten MCT (1998) Ecotoxicology of 3,4-dichloroaniline in enclosed freshwater plankton communities at different nutrient levels. Ecotoxicology 7:49-60.

Jeffries M (1993) Invertebrate colonization of artificial pondweeds of differing fractal dimension. Oikos 67:142-148.

Jenkins DG (1995) Dispersal-limited zooplankton distribution and community composition in new ponds. Hydrobiologia 313/314:15-20.

Johnson ML, Huggins DG, deNoyelles F Jr (1994) Structural equation modeling and ecosystem analysis. In: Graney RL, Kennedy JH, Rodgers JH Jr (eds) Aquatic Mesocosm Studies in Ecological Risk Assessment. Lewis, Boca Raton, pp 627-652.

Kangas P, Adey W (1996) Mesocosms and ecological engineering. Ecol Eng 6:1-5.

Kasai F, Hanazato T (1995a) Effects of the triazine herbicide, simetryn, on freshwater plankton communities in experimental ponds. Environ Pollut 89:197-202.

Kasai F, Hanazato T (1995b) Genetic changes in phytoplankton communities exposed to the herbicide simetryn in outdoor experimental ponds. Arch Environ Contam Toxicol 28:154-160.

Kaushik NK, Stephenson GL, Solomon KR, Day KE (1985) Impact of permethrin on zooplankton communities in limnocorrals. Can J Fish Aquat Sci 42:77-85.

Kaushik NK, Solomon KR, Stephenson GL, Day KE (1986) Use of limnocorrals in evaluating the effects of pesticides on zooplankton communities. In: Cairns J Jr (ed) Community Toxicity Testing. ASTM STP 920. American Society for Testing and Materials, Philadelphia, pp 269-290.

Kedwards TJ, Maund SJ, Chapman PF (1999) Community-level analysis of ecotoxicological field studies. II. Replicated-design studies. Environ Toxicol Chem 18:118-127.

Kennedy JH, Johnson ZB, Johnson PC (1994) Sampling and analysis strategy for biological effects in freshwater field tests. In: Hill IR, Heimbach F, Leeuwangh P, Matthiessen P (eds) Freshwater Field Tests for Hazard Assessment of Chemicals. Lewis, Boca Raton, pp 183-189.

Kennedy JH, Johnson ZB, Wise PD, Johnson PC (1995) Model aquatic ecosystems in ecotoxicological research: considerations of design, implementation, and analysis. In: Hoffman DJ, Rattner BA, Burton GA Jr, Cairns J Jr (eds) Handbook of Ecotoxicology. Lewis, Boca Raton, pp 117–162.

Kennedy JH, Ammann LP, Waller WT, Warren JE, Hosmer AJ, Cairns SH, Johnson PC, Graney RL (1999) Using statisticial power to optimize sensitivity of analysis of variance designs for microcosms and mesocosms. Environ Toxicol Chem 18:113–117.

Kersting K (1994) Functional endpoints in field testing. In: Hill IR, Heimbach F, Leeuwangh P, Matthiessen P (eds) Freshwater Field Tests for Hazard Assessment of Chemicals. Lewis, Boca Raton, pp 57–81.

Kersting K, van der Brink PJ (1997) Effects of the insecticide Dursban®4E (active ingredient chlorpyrifos) in outdoor experimental ditches: responses of ecosystem metabolism. Environ Toxicol Chem 16:251–259.

Kettle WD, deNoyelles F Jr, Heacock BD, Kadoum AM (1987) Diet and reproductive success of bluegill recovered from experimental ponds treated with atrazine. Bull Environ Contam Toxicol 38:47–53.

Kline ER, Figueroa RA, Rodgers JH Jr, Dorn PB (1996) Effects of a nonionic surfactant (C_{14-15} AE-7) on fish survival, growth, and reproduction in outdoor stream mesocosms. Environ Toxicol Chem 15:997–1002.

Kosinski RJ (1989) Artificial streams in ecotoxicological research. In: Boudou A, Ribeyre F (eds) Aquatic Ecotoxicology: Fundamental Concepts and Methodologies, Vol. I. CRC Press, Boca Raton, pp 297–316.

Kreutzweiser DP, Kingsbury PD (1987) Permethrin treatments in Canadian forests. Part 2: Impact on stream invertebrates. Pestic Sci 19:49–60.

Kreutzweiser DP, Thomas DR (1995) Effects of a new molt-inducing insecticide, tebufenozide, on zooplankton communities in lake enclosures. Ecotoxicology 4:307–328.

Lagadic L (1999) Biomarkers in invertebrates. Evaluating the effects of chemicals on populations and communities from biochemical and physiological changes in individuals. In: Peakall DB, Walker CH, Migula P (eds) Biomarkers: A Pragmatic Basis for Remediation of Severe Pollution in Eastern Europe. NATO Science Series 2. Environmental Security, Vol. 54. Kluwer, Dordrecht, pp 153–175.

Lagadic L, Caquet T (1998) Invertebrates in testing of environmental chemicals: are they alternatives? Environ Health Perspect 106(suppl 2):593–611.

Lagadic L, Caquet T, Amiard J-C, Ramade F (1997) Biomarqueurs en Écotoxicologie: Aspects Fondamentaux. Masson, Paris.

Lagadic L, Caquet T, Amiard J-C, Ramade F (1998) Utilisation de Biomarqueurs pour la Surveillance de la Qualité de l'Environnement. Lavoisier Tec & Doc, Paris.

Lalli CM (1990) Enclosed Experimental Marine Ecosystems: A Review and Recommendations. Springer-Verlag, New York.

Langeland A, Koksvik JI, Olsen Y, Reinertsen H (1987) Limnocorral experiments in a eutrophic lake—effects of fish on the planktonic and chemical conditions. Polsk Arch Hydrobiol 34:51–65.

La Point TW (1994) Interpreting the results of agricultural microcosm tests: linking laboratory and experimental field results to predictions of effect in natural ecosystems. In: Hill IR, Heimbach F, Leeuwangh P, Matthiessen P (eds) Freshwater Field Tests for Hazard Assessment of Chemicals. Lewis, Boca Raton, pp 83–94.

Larsen DP, deNoyelles F Jr, Stay JF, Shiroyama T (1986) Comparisons of single-species microcosm and experimental pond responses to atrazine exposure. Environ Toxicol Chem 5:179–190.

Lawton JH (1995) Ecological experiments with model systems. Science 269:328–331.

Lay JP, Peither A, Jüttner I, Weiss K (1993) In situ pond mesocosms for ecotoxicological long-term studies. Chemosphere 26:1137–1150.

Layton RJ, Voshell JR Jr (1991) Colonization of new experimental ponds by benthic macroinvertebrates. Environ Entomol 20:110–117.

Lazzaro X, Drenner RW, Stein RA, Smith JD (1992) Planktivores and plankton dynamics: effects of fish biomass and planktivore type. Can J Fish Aquat Sci 49:1466–1473.

Learner MA, Wiles PR, Pickering JG (1989) The influence of aquatic macrophyte identity on the composition of the chironomid fauna in a former gravel pit in Berkshire, England. Aquat Insects 11:183–191.

Lehtinen K-J (1989) Survival, growth and disease of three-spined stickleback, *Gasterosteus aculeatus* L., brood exposed to bleached kraft mill effluents (BKME) in mesocosms. Ann Zool Fenn 26:133–143.

Liber K (1994) Use of limnocorrals for assessing the aquatic fate and effects of pesticides. In: Hill IR, Heimbach F, Leeuwangh P, Matthiessen P (eds) Freshwater Field Tests for Hazard Assessment of Chemicals. Lewis, Boca Raton, pp 199–214.

Liber K, Solomon KR, Carey JH (1997) Persistence and fate of 2,3,4,6-tetrachlorophenol and pentachlorophenol in limnocorrals. Environ Toxicol Chem 16:293–305.

Liber K, Kaushik NK, Solomon KR, Carey JH (1992) Experimental designs for aquatic mesocosm studies: a comparison of the ANOVA and regression design for assessing the impact of tetrachlorophenol on zooplankton populations in limnocorrals. Environ Toxicol Chem 11:61–77.

Lowe RL, Guckert JB, Belanger SE, Davidson DH, Johnson DW (1996) An evaluation of periphyton community structure and function on tile and cobble substrata in experimental stream mesocosms. Hydrobiologia 328:135–146.

Lozano SL, O'Halloran SL, Sargent KW, Brazner JC (1992) Effects of esfenvalerate on aquatic organisms in littoral enclosures. Environ Toxicol Chem 11:35–47.

Lynch TR, Johnson HE, Adams WJ (1985) Impact of atrazine and hexachlorobiphenyl on the structure and function of model stream ecosystems. Environ Toxicol Chem 4:399–413.

Mackay D (1989) An approach to modelling the long term behavior of an organic contaminant in a large lake: application to PCBs in Lake Ontario. J Great Lakes Res 15:283–297.

Mackay D (1990) Atmospheric contributions to contamination of Lake Ontario. In: Kurtz DA (ed) Long Range Transport of Pesticides. Lewis, Chelsea, pp 317–328.

Mackay D, Joy M, Paterson S (1983) A quantitative water, air, sediment interaction (QWASI) fugacity model for describing the fate of chemicals in lakes. Chemosphere 12:981–997.

Maguire RJ, Carey JH, Hart JH, Tkacz RJ, Lee HB (1989) Persistence and fate of deltamethrin sprayed on a pond. J Agric Food Chem 37:1153–1159.

Maltby L (1992) The use of the physiological energetics of *Gammarus pulex* to assess toxicity: a study using artificial streams. Environ Toxicol Chem 11:79–85.

Maltby L (1994) Stress, shredders and streams: using *Gammarus* energetics to assess water quality. In: Sutcliffe DW (ed) Water Quality and Stress Indicators in Marine and Freshwater Ecosystems: Linking Levels of Organisation (Individuals, Populations, Communities). Freshwater Biological Association, Ambleside, pp 98–110.

Matthiessen P (1994) Guidelines for pesticide hazard assessment with freshwater field tests at the international level. In: Hill IR, Heimbach F, Leeuwangh P, Matthiessen P (eds) Freshwater Field Tests for Hazard Assessment of Chemicals. Lewis, Boca Raton, pp 455–473.

Mayasich JM, Kennedy JH, O'Grodnick JS (1994) Evaluation of the ecological and biological effects of tralomethrin utilizing an experimental pond system. In: Graney RL, Kennedy JH, Rodgers JH Jr (eds) Aquatic Mesocosm Studies in Ecological Risk Assessment. Lewis, Boca Raton, pp 497–515.

Mazumder A, McQueen DJ, Taylor WD, Lean DRS (1988) Effects of fertilization and planktivorous fish (yellow perch) predation on size distribution of particulate phosphorus and assimilated phosphate: large enclosures experiments. Limnol Oceanogr 33:421–430.

McCarthy JF, Shugart LR (1990) Biomarkers of Environmental Contamination. Lewis, Boca Raton.

Menzel DW, Case J (1977) Concept and design: controlled ecosystem pollution experiment. Bull Mar Sci 27:1–7.

Mitchell GC (1994) System design for field tests in flowing waters: design and use of outdoor artificial streams in ecotoxicology. In: Hill IR, Heimbach F, Leeuwangh P, Matthiessen P (eds) Freshwater Field Tests for Hazard Assessment of Chemicals. Lewis, Boca Raton, pp 127–139.

Mitchell GC, Bennett D, Pearson N (1993) Effects of lindane on macroinvertebrates and periphyton in outdoor artificial streams. Ecotoxicol Environ Saf 25:90–102.

Morris RG, Kennedy JH, Johnson PC, Hambleton FE (1994) Pyrethroid insecticide effects on bluegill sunfish in microcosms and mesocosms and bluegill impact on microcosm fauna. In: Graney RL, Kennedy JH, Rodgers JH Jr (eds) Aquatic Mesocosm Studies in Ecological Risk Assessment. Lewis, Boca Raton, pp 373–395.

Nabholz JV, Clements RG, Zeeman MG (1997) Information needs for risk assessment in EPA's Office of Pollution Prevention and Toxics. Ecol Appl 7:1094–1098.

Newman RM (1991) Herbivory and detritivory on freshwater macrophytes by invertebrates: a review. J North Am Benthol Soc 10:89–114.

Odum EP (1984) The mesocosm. BioScience 34:558–562.

Odum HT (1996) Scales of ecological engineering. Ecol Eng 6:7–19.

OECD (1996) Guidelines for Testing of Chemicals. Draft Proposal for a Guidance Document. Freshwater Lentic Field Tests. Organization for Economic Cooperation and Development, Paris.

O'Halloran SL, Liber K, Schmude KL, Corry TD (1996) Effects of diflubenzuron on benthic macroinvertebrates in littoral enclosures. Arch Environ Contam Toxicol 30: 444–451.

O'Neil PE, Harris SC, Mettee MF (1990) Experimental stream mesocosms as applied in the assessment of produced water effluents associated with the development of coalbed methane. In: Cuffney TF (ed) Experimental Ecosystems: Applications to Ecotoxicology. Technical Information Workshop, North American Benthological Society, Virginia Polytechnic Institute and State University, Blacksburg, VA.

Oviatt CA, Quinn JG, Maughan JT, Ellis JT, Sullivan BK, Gearing JN, Gearing PJ, Hunt CD, Sampou PA, Latimer JS (1987) Fate and effects of sewage sludge in the coastal marine environment: a mesocosm experiment. Mar Ecol Prog Ser 41:187–203.

Paterson CG, Fernando CH (1970) Benthic fauna colonization of a new reservoir with particular reference to the Chironomidae. J Fish Res Board Can 27:213–232.

Peakall DB (1992) Animal Biomarkers as Pollution Indicators. Chapman & Hall, London.

Perschbacher PW, Stone N, Ludwig GM, Guy CB Jr (1996) Evaluation of effects of common aerially-applied soybean herbicides and propanil on the plankton communities of aquaculture ponds. Aquaculture 157:117–122.

Pignatello JJ, Martinson MM, Steirt JG, Carlson RE, Crawford RL (1983) Biodegradation and photolysis of pentachlorophenol in artificial freshwater streams. Appl Environ Microbiol 46:1024–1031.

Pusey BJ, Arthington AH, Flanders TJ (1994a) An outdoor replicated artificial stream system: design, operating conditions and initial invertebrate colonisation. Ecotoxicol Environ Saf 27:177–191.

Pusey BJ, Arthington AH, McLean J (1994b) The effects of a pulsed application of chlorpyrifos on macroinvertebrate communities in an outdoor artificial stream system. Ecotoxicol Environ Saf 27:221–250.

Qin J, Culver DA (1996) Effect of larval fish and nutrient enrichment on plankton dynamics in experimental ponds. Hydrobiologia 321:109–118.

Ratte H-R, Poethke HJ, Dülmer U, Hommen U (1994) Modelling of aquatic field tests for hazard assessment. In: Hill IR, Heimbach F, Leeuwangh P, Matthiessen P (eds) Freshwater Field Tests for Hazard Assessment of Chemicals. Lewis, Boca Raton, pp 399–423.

Ravera O (1989) The enclosure method: concepts, technology, and some examples of experiments with trace metals. In: Boudou A, Ribeyre F (eds) Aquatic Ecotoxicology: Fundamental Concepts and Methodologies, Vol. I. CRC Press, Boca Raton, pp 249–272.

Regal RR, Lozano SJ (1994) Optimal design of aquatic field studies. In: Graney RL, Kennedy JH, Rodgers JH Jr (eds) Aquatic Mesocosm Studies in Ecological Risk Assessment. Lewis, Boca Raton, pp 157–171.

Reinertsen H, Jensen A, Kokvsik JI, Langeland A, Olsen Y (1990) Effects of fish removal on the limnetic ecosystem of a eutrophic lake. Can J Fish Aquat Sci 47:166–173.

Resetarits WJ (1991) Ecological interactions among predators in experimental stream communities. Ecology 72:1782–1793.

RESOLVE (1992) Improving aquatic risk assessment under FIFRA. Report of the Aquatic Effects Dialogue Group. Resolve, World Wildlife Fund, Washington, DC.

Rodgers JH Jr (1994) Design of mesocosm studies and statistical analysis of data. Summary and discussion. In: Graney RL, Kennedy JH, Rodgers JH Jr (eds) Aquatic Mesocosm Studies in Ecological Risk Assessment. Lewis, Boca Raton, pp 173–175.

Rodgers JH Jr, Crossland NO, Kline ER, Gillespie WB, Figueroa RA, Dorn PB (1996) Design and construction of model stream ecosystems. Ecotoxicol Environ Saf 33:30–37.

Ronday R, Aalderink GH, Crum SJH (1998) Application methods of pesticides to an aquatic mesocosm in order to simulate effects of spray drift. Water Res 32:147–153.

Rosenzweig MS, Buikema AL Jr (1994) Phytoplankton colonization and seasonal succession in new experimental ponds. Environ Toxicol Chem 13:599–605.

Schramm KW, Goss KU (1990) MESIP—modelling environmental scenarios in ponds. Toxicol Environ Chem 26:123–128.

Schramm HL Jr, Jirka KJ (1989) Epiphytic macroinvertebrates as a food resource for bluegills in Florida lakes. Trans Am Fish Soc 118:416–426.

Schramm HL Jr, Jirka KJ, Hoyer MV (1987) Epiphytic macroinvertebrates on dominant macrophytes in two central Florida lakes. J Freshwater Ecol 4:151–161.

Schüürmann G (1998) Ecotoxic modes of action of chemical sustances. In: Schüürmann G, Markert B (eds) Ecotoxicology. Ecological Fundamentals, Chemical Exposure and

Biological Effects. Wiley, New York (Spektrum Academischer Verlag, Heidelberg), pp 665–750.

SETAC-Europe (1992) Guidance Document on Testing Procedures for Pesticides in Freshwater Mesocosms. SETAC-Europe, Brussels.

SETAC-RESOLVE (1992) Guidance Document on Testing Procedures for Pesticides in Freshwater Static Mesocosms. Report of a meeting held at Wintergreen, USA, October 11, 1991.

Shaw JL, Kennedy JH (1996) The use of aquatic field mesocosm studies in risk assessment. Environ Toxicol Chem 15:605–607.

Shaw JL, Manning JP (1996) Evaluating macroinvertebrate population and community level effects in outdoor microcosms: use of in situ bioassays and multivariate analysis. Environ Toxicol Chem 15:608–617.

Shaw JL, Moore M, Kennedy JH, Hill IR (1994) Design and statistical analysis of field aquatic mesocosm studies. In: Graney RL, Kennedy JH, Rodgers JH Jr (eds) Aquatic Mesocosm Studies in Ecological Risk Assessment. Lewis, Boca Raton, pp 85–103.

Shaw JL, Maund SJ, Hill IR (1995a) Fathead minnow reproduction in outdoor microcosms: a comparison to bluegill sunfish reproduction in large mesocosms. Environ Toxicol Chem 14:1753–1762.

Shaw JL, Maund SJ, Marshall SJ, Hill IR (1995b) Fathead minnow (*Pimephales promelas* Rafinesque) reproduction in outdoor microcosms: an assessment of the ecological effects of fish density. Environ Toxicol Chem 14:1763–1772.

Shaw JL, Copeland JR, Johnson JK (1995c) Use of coded wire tag to identify fathead minnow (*Pimephales promelas* Rafinesque) adults in an outdoor microcosm study designed to evaluate consistency in reproduction. Environ Toxicol Chem 14:1773–1780.

Shires SW (1983) The use of small enclosures to assess the toxic effects of cypermethrin on fish under field conditions. Pestic Sci 14:475–480.

Siefert RE, Lozano SJ, Brazner JC, Knuth ML (1989) Littoral enclosures for aquatic field testing of pesticides: effects of chlorpyrifos on a natural system. In: Voshell JR Jr (ed) Using Mesocosms to Assess the Aquatic Ecological Risk of Pesticides. Misc Publ Entomol Soc Am 75:57–73.

Siefert RE, Lozano SJ, Knuth ML, Heinis LJ, Brazner JC, Tanner DK (1990) Pesticide testing with littoral enclosures. In: Cuffney TF (ed) Experimental Ecosystems: Applications to Ecotoxicology. Technical Information Workshop, North American Benthological Society, Virginia Polytechnic Institute and State University, Blacksburg, VA.

Sierszen ME, Lozano SJ (1998) Zooplankton population and community responses to the pesticide azinphos-methyl in freshwater littoral enclosures. Environ Toxicol Chem 17:907–914.

Solomon KR, Yoo JY, Lean D, Kaushik NK, Day KE, Stephenson GL (1985) Dissipation of permethrin in limnocorrals. Can J Fish Aquat Sci 42:70–76.

Solomon KR, Yoo JY, Lean D, Kaushik NK, Day KE, Stephenson GL (1986) Methoxychlor distribution, dissipation, and effects in freshwater limnocorrals. Environ Toxicol Chem 5:577–586.

Solomon KR, Stephenson GL, Kaushik NK (1989) Effects of methoxychlor on zooplankton in freshwater enclosures: influence of enclosure size and number of applications. Environ Toxicol Chem 8:659–669.

Sörensson F, Pettersson K, Semer J-H, Sahlsten E (1989) Flows of nitrogen in a mesocosm experiment in the Baltic Sea. Mar Ecol Prog Ser 58:77–88.

Southwood TRE (1978) Ecological Effects, 2nd ed. Chapman & Hall, New York.

Sparks TH, Scott WA, Clarke RT (1999) Traditional multivariate techniques: potential for use in ecotoxicology. Environ Toxicol Chem 18:118–127.

Sparling DW, Lowe TP (1996) Metal concentrations of tadpoles in experimental ponds. Environ Pollut 91:149–159.

Stay FS, Jarvinen AW (1995) Use of microcosm and fish toxicity data to select mesocosm treatment concentrations. Arch Environ Contam Toxicol 28:451–458.

Stephenson GL, Hamilton P, Kaushik NK, Robinson JB, Solomon KR (1984) Spatial distribution of plankton in enclosures of three sizes. Can J Fish Aquat Sci 41:1048–1054.

Stephenson GL, Kaushik NK, Solomon KR, Day KE (1986) Impact of methoxychlor on freshwater communities of plankton in limnocorrals. Environ Toxicol Chem 5: 587–603.

Stephenson M, Mackie GL (1986) Effects of 2,4-D treatment on natural benthic macroinvertebrate communities in replicate artificial ponds. Aquat Toxicol 9:243–251.

Stephenson RR, Kane DF (1984) Persistence and effects of chemicals in small enclosures in ponds. Arch Environ Contam Toxicol 13:313–326.

Stokes W, Marafante E, Peakall D, Goldstein B (eds) (1998) Alternative testing methodologies. Environ Health Perspect 106(suppl 2):401–620.

Stunkard CL (1994) Tests of proportional means for mesocosm studies. In: Graney RL, Kennedy JH, Rodgers JH Jr (eds) Aquatic Mesocosm Studies in Ecological Risk Assessment. Lewis, Boca Raton, pp 71–83.

Street M, Titmus G (1979) The colonization of experimental ponds by Chironomidae (Diptera). Aquat Insects 1:233–244.

Sugiura K, Aoki M, Kaneko S, Daisaku I, Komatsu Y, Shibuya H, Suzuki H, Gogo M (1984) Fate of 2,4,6-trichlorophenol, pentachlorophenol, p-chlorobiphenyl, and hexachlorobenzene in an outdoor experimental pond: comparison between observations and predictions based on laboratory data. Arch Environ Contam Toxicol 13:745–758.

Suter GW II (1995) Introduction to ecological risk assessment for aquatic toxic effects. In: Rand GM (ed) Fundamentals of Aquatic Toxicology, 2nd Ed. Taylor & Francis, Washington, DC, pp 803–816.

Swift MC, Troelstrup HN Jr, Detenbeck NE, Foley JL (1993) Large artificial streams in toxicological and ecological research. J North Am Benthol Soc 12:359–366.

Takamura K (1995) Chironomids fail to emerge from LAS-contaminated water. Ecotoxicology 4:245–257.

Tana J, Rosemarin A, Lehtinen K-J, Härdig J, Grahn O, Landner L (1994) Assessing impacts on Baltic coastal ecosystems with mesocosm and fish biomarker tests: a comparison of new and old wood pulp bleaching technologies. Sci Total Environ 145: 213–234.

Tanner DK, Knuth ML (1995) Effects of azinphos-methyl on the reproductive success of the bluegill sunfish, *Lepomis macrochirus*, in littoral enclosures. Ecotoxicol Environ Saf 32:184–193.

Taylor EJ, Maund SJ, Bennett D, Pascoe D (1994) Effects of 3,4-dichloroaniline on the growth of two freshwater macroinvertebrates in a stream mesocosm. Ecotoxicol Environ Saf 29:80–85.

Thompson DG, Holmes SB, Pitt DG, Solomon KR, Wainio-Keizer KL (1994) Applying concentration-response theory to experimental design of aquatic enclosure studies. In: Graney RL, Kennedy JH, Rodgers JH Jr (eds) Aquatic Mesocosm Studies in Ecological Risk Assessment. Lewis, Boca Raton, pp 129–156.

Thorp JH, Delong MD (1994) The riverine productivity model: an heuristic view of carbon sources and organic processing in large rivers ecosystems. Oikos 70:305–308.

Threlkeld ST, Soballe DM (1988) Effects of mineral turbidity on fresh water plankton communities: three exploratory tank experiments of factorial design. Hydrobiologia 159:223–236.

Touart LW (1988) Aquatic Mesocosm Test to Support Pesticide Registrations. Hazard Evaluation Division Technical Guidance Document. EPA/540/09-88-035. U.S. E.P.A., Washington, D.C.

Touart LW (1994) Regulatory endpoints and the experimental design of aquatic mesocosm tests. In: Graney RL, Kennedy JH, Rodgers JH Jr (eds) Aquatic Mesocosm Studies in Ecological Risk Assessment. Lewis, Boca Raton, pp 25–33.

Touart LW, Maciorowski AF (1997) Information needs for pesticide registration in the United States. Ecol Appl 7:1086–1093.

USEPA-OPPTS (1996) Ecological Effects Test Guidelines. OPPTS 850.1950. Field Testing for Aquatic Organisms. EPA 712-C-96-135. U.S. E.P.A., Washington, DC.

van den Brink PJ, Ter Braak CJF (1999) Principal response curves: analysis of time-dependent multivariate responses of biological community to stress. Environ Toxicol Chem 18:138–157.

van den Brink PJ, Wijngaarden RPA, Lucassen WGH, Brock TCM, Leeuwangh P (1996) Effects of the insecticide Dursban® 4E (active ingredient chlorpyrifos) in outdoor experimental ditches: II. Invertebrate community responses and recovery. Environ Toxicol Chem 15:1143–1153.

van Leeuwen K, Emans H-J, Van de Plassche E, Canton H (1994) The role of field tests in hazard assessment of chemicals. In: Hill IR, Heimbach F, Leeuwangh P, Matthiessen P (eds) Freshwater Field Tests for Hazard Assessment of Chemicals. Lewis, Boca Raton, pp 425–437.

van Wijngaarden RPA, van den Brink PJ, Crum SJH, Oude Voshaar JH, Brock TCM, Leeuwangh P (1996) Effects of the insecticide Dursban® 4E (active ingredient chlorpyrifos) in outdoor experimental ditches: I. Comparison of short-term toxicity between the laboratory and the field. Environ Toxicol Chem 15:1133–1142.

Vanni MJ, Layne CD (1997) Nutrient recycling and herbivory as mechanisms in the top-down effect of fish on algae in lakes. Ecology 78:21–40.

Vanni MJ, Layne CD, Arnott SE (1997) Top-down trophic interactions in lakes: effects of fish on nutrient dynamics. Ecology 78:1–20.

Vannote RL, Minshall GW, Cummins KW, Sedell JR, Cushing CE (1980) The river continuum concept. Can J Fish Aquat Sci 37:130–137.

Verdonschot PFM (1996) Oligochaetes and eutrophication; an experiment over four years in outdoor mesocosms. Hydrobiologia 334:169–183.

Verdonschot PFM, Ter Braak CJF (1994) An experimental manipulation of oligochaete communities in mesocosms treated with chlorpyrifos or nutrient additions: multivariate analyses with Monte Carlo permutation tests. Hydrobiologia 278:251–266.

Vethaak AD, Jol JG, Meijboom A, Eggen ML, Rheinallt T, Wester PW, van de Zande T, Bergman A, Dankers N, Ariese F, Baan RA, Everts JM, Opperhuizen A, Marquenie JM (1996) Skin and liver diseases induced in flounder (*Platichthys flesus*) after long-term exposure to contaminated sediments in large-scale mesocosms. Environ Health Perspect 104:1218–1229.

Vinyard GL, Drenner RW, Gophen M, Pollingher U, Winkleman DL, Hambright KD (1988) An experimental study of the plankton community impacts of two omnivorous filter-feeding cichlids, *Tilapia galilaea* and *Tilapia aurea*. Can J Fish Aquat Sci 45: 685–690.

Voshell JR Jr (1990) Introduction and overview of mesocosms. In: Cuffney TF (ed)

Experimental Ecosystems: Applications to Ecotoxicology. Technical Information Workshop, North American Benthological Society, Virginia Polytechnic Institute and State University, Blacksburg, VA.

Voshell JR Jr, Simmons GM Jr (1984) Colonization and succession of benthic macroinvertebrates in a new reservoir. Hydrobiologia 112:27–39.

Wakeham SG, Davis AC, Karas JA (1983) Mesocosm experiments to determine the fate and persistence of volatile organic compounds in coastal seawater. Environ Sci Technol 17:611–617.

Walker CH (1998) Biomarker strategies to evaluate the environmental effects of chemicals. Environ Health Perspect 106(suppl 2):613–620.

Walker CH, Kaiser K, Klein W, Lagadic L, Peakall DB, Sheffield SR, Soldan T, Yasuno M (1998) Alternative testing methodologies for ecotoxicity. Environ Health Perspect 106(suppl 2):441–451.

Wallace JB, Vogel DS, Caffney TF (1986) Recovery of a headwater stream from an insecticide-induced community disturbance. J North Am Benthol Soc 5:115–126.

Wallace JB, Caffney TF, Lay CC, Vogel D (1987) The influence of an ecosystem-level manipulation on prey consumption by a lotic dragonfly. Can J Zool 65:35–40.

Wangberg S-Å, Heyman U, Blanck H (1991) Long-term and short-term arsenate toxicity to freshwater phytoplankton and periphyton in limnocorrals. Can J Fish Aquat Sci 48:173–182.

Ward S, Arthington AH, Pusey BJ (1995) The effects of a chronic application of chlorpyrifos on the macroinvertebrate fauna in an outdoor artificial stream system: species responses. Ecotoxicol Environ Saf 30:2–23.

Wayland M (1991) Effect of carbofuran on selected macroinvertebrates in a prairie parkland pond: an enclosure approach. Arch Environ Contam Toxicol 21:270–280.

Wayland M, Boag DA (1990) Toxicity of carbofuran to selected macroinvertebrates in prairie ponds. Bull Environ Contam Toxicol 45:74–81.

Weber EC, Deutsch WG, Bayne DR, Seesock WC (1992) Ecosystem-level testing of a synthetic pyrethroid insecticide in aquatic mesocosms. Environ Toxicol Chem 11:87–105.

Whittle D, Wong DCL, Dorn PB, Maltby L, Tattersfield LJ, Warren P (1999) Long term spatial and temporal variability in invertebrate community structure in model ecosystems. Abstract n°1f/P008, 9th Annual Meeting of SETAC-Europe, 25–29 May 1999, Leipzig.

Williams JR, Dyke PT, Fuchs WW, Benson VW, Rice OW, Taylor ED (1990) EPIC— Erosion/Productivity Impact Calculator: 2. User Manual. Tech Bull 1768. U.S. Department of Agriculture, Washington, DC.

Wolff CJM, Crossland NO (1985) Fate and effects of 3,4-dichloroaniline in the laboratory and in outdoor ponds. I. Fate. Environ Toxicol Chem 4:481–487.

Yasuno M, Hanazato T, Iwakuna T, Takamura K, Ueno R, Takamura N (1988) Effects of permethrin on phytoplankton and zooplankton in an enclosure ecosystem in a pond. Hydrobiologia 159:247–258.

Zischke JA, Arthur JW, Hermanutz RO, Hedtke SF, Helgen JC (1985) Effects of pentachlorophenol on invertebrates and fish in outdoor experimental channels. Aquat Toxicol 7:37–58.

Manuscript received January 23, 1999; accepted July 28, 1999.

Lizard Contaminant Data for Ecological Risk Assessment

Kym Rouse Campbell·Todd S. Campbell

Contents

I. Introduction	39
II. Previous Reviews	42
III. Review Methods	43
IV. Residue or Bioaccumulation Data	43
A. Organic Contaminants	59
B. Inorganic Contaminants	72
V. Lethal and Sublethal Effects Data	73
A. Organic Contaminants	91
B. Inorganic Contaminants	104
C. Radionuclides/Radiation	105
VI. Discussion	108
Summary	111
Acknowledgments	112
References	112

I. Introduction

Ecological risk assessments, to be realistic, should include a full complement of the relevant members of the systems being studied. Reptiles are important constituents and comprise a large percentage of the faunal biomass in many terrestrial and aquatic ecosystems. They are predators and prey of vertebrates and invertebrates, and their unique life histories make their roles in food webs diverse and important. In addition, reptiles are crucial to the proper functioning of many ecological processes. However, reptiles are rarely included in ecological risk assessments because either contaminant data are not available or they are not considered to be important in ecosystem functions. Certainly, the former lack has been a direct result of the latter opinion. Existing risk assessment schemes lose their predictive value when important taxa, such as reptiles, are missing, especially in risk assessments performed for terrestrial arid ecosystems (van der Valk 1997). Reptiles also are infrequently considered in habitat evalua-

---

Communicated by George W. Ware

K. R. Campbell (✉)
The Cadmus Group, Inc.,
136 Mitchell Road, Oak Ridge, TN 37830, U.S.A.

T. S. Campbell
Department of Ecology and Evolutionary Biology, University of Tennessee,
569 Dabney Hall, Knoxville, TN 37996-1610, U.S.A.

tion and management (Fontenot et al. 1996). By neglecting reptiles, we evaluate only a portion of the biotic community and cannot fully assess the risks posed by human intervention.

Reptiles are more sensitive to the effects of pollutants, such as organochlorine pesticides, than are birds or mammals, but reptiles are apparently less sensitive than fish (Hall 1980). Because reptiles are ectothermic, have a low metabolic rate, and have relatively simple enzyme systems, they have a poor ability to quickly detoxify pesticides that are absorbed, inhaled, or ingested (Walker and Ronis 1989). Toxic Equivalency Factors developed for use in risk assessments for fish, birds, and humans are not predictive of toxicity in reptiles (Bishop and Gendron 1998). In fact, the impacts of pollutants on reptiles have rarely been quantified and monitored in the wild (Bishop and Gendron 1998), and the toxicity of most chemicals to reptiles in drylands is virtually unknown (Everts 1997).

The value of reptiles as indicators of local contamination has been recognized (Hall 1980; Meyers-Schöne and Walton 1994; Bishop and Gendron 1998). Because many adult reptiles are secondary and tertiary predators, they are susceptible to bioaccumulation of persistent environmental contaminants (Bishop and Gendron 1998), especially organochlorine pesticides (Lambert 1993, 1997a,b). In addition, reptiles are effective biomonitors of heavy metal (Overmann and Krajicek 1995) and radionuclide (Meyers-Schöne et al. 1993) contamination. They bioaccumulate and biomagnify contaminants to levels equal to or greater than those reported for birds and mammals (Olafsson et al. 1983; Bryan et al. 1987; Hall and Henry 1992). Recently, reptiles have been used as biomonitors of another class of environmental contaminants: endocrine disrupters (Crain and Guillette 1998).

Of the approximately 6,000 species of reptiles, more than half are lizards (order Squamata, suborder Lacertilia), an extremely diverse, cosmopolitan group of terrestrial, fossorial, and arboreal predators and herbivores. Twenty-six families of lizards are currently recognized within the suborder Lacertilia (Pough et al. 1998). Lizards are particularly well represented in arid and tropical regions, but are important constituents of many temperate ecosystems as well. In desert ecosystems of the world, lizards make up a large part of vertebrate density and diversity, and most are insectivorous. For example, many hundreds of species of skinks (Scincidae) have been described from the arid regions of Australia and Africa and reach extraordinary densities (Pianka 1986). In the southwestern United States, insectivorous lizard faunas are also diverse, although less so than Old World arid lizard faunas (Pianka 1986; Wright and Vitt 1993). In the New World tropics and the Caribbean, well over 500 species are represented from just two lizard genera: *Sphaerodactylus* (Gekkonidae) and *Anolis* (Polychrotidae), in addition to numerous other lizard groups (Schwartz and Henderson 1991). On many islands in the Caribbean, in Central and South America, and in Florida, *Anolis* lizards assume roles normally played by birds and can attain densities as high as $1/m^2$, and thus much of the carbon and energy flow through their populations (Roughgarden 1995). Recently, *Sphaerodactylus* were found at densities greater than 50,000/ha (Rodda et al. 1999), the highest density of any terrestrial vertebrate. These small terrestrial and arboreal insectivorous lizards occupy intermediate trophic levels and effectively convert large numbers of arthropods into body biomass that in turn becomes available to larger predators in higher trophic levels. The substantially larger monitor lizards (Varanidae) of Old World arid and tropical zones assume the roles of top predators (Pianka 1986), and the large New World iguanids (Iguanidae) are generally herbivores and represent primary consumers (Schwartz and Henderson 1991).

Lizards are important components of terrestrial ecosystems as well as integral components of food webs; they form an important link in food chains between invertebrate prey and predatory vertebrates such as birds and snakes (Lambert 1997a,b). Contaminants that find their way into the environment could even reach humans through bioconcentration pathways that include lizards. However, lizards have been either ignored or represented as incidental components of contaminant studies. They have rarely been used as bioindicators of pollution for various reasons, including difficulty in sampling large enough numbers, ancient fear, and their minor economic importance (Loumbourdis 1997). In contrast to birds and mammals, lizards are good indicators of the quality of terrestrial habitats because of their relatively restricted mobility (Lambert 1993). In addition, lizards are sensitive to many pesticides as the result of direct exposure or intake via their prey (Lambert 1993).

Many lizards are listed as threatened and endangered species. The U.S. Fish and Wildlife Service list of threatened and endangered species for the U.S. includes 9 species of lizards as of October 31, 1999; 4 species are listed as endangered, and 5 species are listed as threatened (http://www.fws.gov). The Committee on the Status of Endangered Wildlife in Canada lists 3 species of lizards as vulnerable; no species of lizards is currently listed as threatened or endangered in Canada (http://www.mcgill.ca/redpath/cosehome.htm). Environment Australia includes 7 species of lizards on their list of endangered reptiles and 24 species of lizards on their list of vulnerable reptiles (as of September 1998) (http://www.biodiversity.environment.gov.au/plants/threaten/lists/anzeec_lists/anzfauna.htm). In addition, the International Union for the Protection of Nature (IUCN) or World Conservation Union 1996 Red List of Threatened Animals includes 139 lizard species (http://www.wcmc.org.uk); 20 lizard species are listed as critical, 22 are listed as endangered, 74 species are listed as vulnerable, and 23 are listed as near threatened or conservation dependent.

We propose that representative species of lizards should be included in ecological risk assessments, especially for those studies done in areas where lizards are abundant and diverse and thus play major roles in ecosystem processes. We review and summarize environmental contaminant residue, lethal effects, and sublethal effects data for lizards for four reasons: (1) to stress the importance of lizards and encourage their inclusion in ecological risk assessments; (2) to demonstrate the paucity of available contaminant data on lizards and reveal the main information gaps; (3) to encourage further ecotoxicological studies on lizards; and (4) to facilitate the use of existing lizard contaminant data in ecological risk assessments.

II. Previous Reviews

Hall (1980) reviewed the effects of environmental contaminants on reptiles and included a few studies on lizards. His review is concerned almost entirely with organochlorine pesticides. Many of the data are based on observations following field applications or on reports of chemical residues in reptile tissues. Hall (1980) proposed that the following generalizations may apply to reptiles: (1) sensitivity is species specific; (2) species at higher trophic levels are most affected; (3) effects of contaminants vary considerably, depending on the physiological state of the animal; (4) toxicity depends on the form of the contaminants or its degradates to which the reptile is exposed; (5) specific enzyme systems are affected, producing a variety of sublethal effects; and (6) long-lived species are more susceptible than short-lived species. Hall suggested that the efforts devoted to contaminant effects on birds, mammals, and aquatic organisms should be duplicated on reptiles. Even though 20 years have passed since Hall's review (1980), most of the questions he discussed still have not been answered.

Years later, Hall and Henry (1992) reviewed the status and needs of assessing the effects of pesticides on reptiles, including lizards. They stated that measures to conserve reptiles have been slow to materialize and that threats from toxic chemicals in the environment have received insufficient attention. Almost no experimental evaluations of the sensitivity of reptiles to environmental chemicals have been made, although field studies were common in the era of organochlorines (Hall and Henry 1992). They concluded that far too little is known to conclude that safety standards for other kinds of vertebrates are adequate for the protection of reptiles.

Lambert (1997a) recently thoroughly reviewed the effects of pesticides on reptiles in sub-Saharan Africa; the review includes many studies on lizards. Most of the reviewed studies relate to deaths from organochlorine insecticides sprayed for the control of tsetse flies (*Glossina* spp.). He suggests that pesticide use in Africa is likely to increase substantially in the next decade or so and that future work is required regarding the threat of pesticides to reptile species diversity. Pesticide usage is a reflection of changed land use when woodland and other natural vegetation is cleared for agriculture, and a substantial change in land use is likely to have a greater impact on populations of reptiles than the direct effects of pesticides themselves. Lizards possess certain characteristics that make them suitable candidates as bioindicators, and a list of characteristics is presented in the review (Lambert 1997a).

Data from Hall's (1980) and Lambert's (1997a) reviews were not included in this review. We suggest consulting those reviews in addition to the information presented here. To our knowledge, our report represents the most comprehensive review of lizard contaminant data undertaken to date. We review here the bioaccumulation and effects of organic and inorganic contaminants and radionuclides/radiation on lizards because they are usually the contaminants of concern in ecological risk assessments.

III. Review Methods

Available organic and inorganic contaminant residue or bioaccumulation, lethal effects, and sublethal effects studies on lizards as a result of detailed literature reviews and database searches were obtained. Studies on substances that affect sex determination in lizards were included because of their applicability regarding the effects of environmental estrogens or endocrine-disrupting contaminants on reptile reproduction. In addition, available studies were obtained concerning the accumulation and effects of radionuclides/radiation on lizards. Studies on substances that were not considered environmental contaminants (i.e., drugs, dyes) are not included in this review.

Data from studies that were not to genus or species level (i.e., "lizards") were excluded. If the genus, species, or common name of the lizard had changed since the time of study publication, the currently accepted scientific and common names of lizards according to Stebbins (1985), Conant and Collins (1991), and Frank and Ramus (1995) have been used. If data from multiple individuals were available, the mean and range were calculated or presented. A knowledge of the range of concentration of a particular contaminant is important in ecological risk assessments. The detection limit, if available, was indicated and used when calculating the mean. Data regarding background and low levels of contaminants, even zero or below-detection limits, are important; therefore, when available, they also were included. Summary tables were arranged for each class of contaminants (organics, inorganics, and radionuclides) in chronological order from oldest to the most recent study and were designed to include information useful for ecological risk assessments.

Many methods of contaminant concentration analysis were used in the reviewed studies, and analytical methods were not reviewed. Because analytical methods have improved over time and detection limits have decreased, more recent data may be more reliable than data from older studies. Because of the many methods used, concentrations between locations or years were not compared. Consulting the individual studies for information regarding the analytical methods is suggested. Because concentrations were not compared, the units reported were not standardized. However, most of the concentrations are in parts per million (ppm); a few are reported in parts per billion (ppb). If available, wet or dry weight was indicated. Where necessary, application rates, sizes, and distances were converted to metric units. In addition, the quality of the study was not considered.

IV. Residue or Bioaccumulation Data

Lizard contaminant studies that contained residue or bioaccumulation data were available for species of 11 families of lizards (Table 1). The majority of contaminant residue data available for lizards concerned organic contaminants (seven studies) (Table 2). Inorganic contaminant concentrations in lizards were avail-

Table 1. Number of Reviewed Lizard Contaminant Studies by Family (Pough et al. 1998). The Number of Species in each Family in Reviewed Studies are in Parentheses.

Family	Residue or bioaccumulation studies		Lethal effects studies			Sublethal effects studies		
	Organics	Inorganics	Organics	Inorganics	Radionuclides/radiation	Organics	Inorganics	Radionuclides/radiation
Agamidae (agamas)	1 (1)	1 (1)	1 (1)		1 (1)	1 (1)	1 (1)	1 (1)
Chamaeleondae (chameleons)								
Iguanidae (iguanas and spinytail iguanas)		1 (2)						
Opluridae (Madagascar iguanas)								
Phrynosomatidae (earless, spiny, tree, side-blotched, and horned lizards)	1 (1)				4 (1)			15 (4)
Tropiduridae (Neotropical ground lizards)								
Polychrotidae (anoles)	1 (1)		1 (1)		2 (3)	1 (1)		1 (2)
Hoplocercidae (dwarf iguanas, lesser spinytail iguanas, and weapontails)								
Crotaphytidae (collared and leopard lizards)					1 (1)			6 (1)

Table 1. (Continued).

Family	Residue or bioaccumulation studies		Lethal effects studies			Sublethal effects studies		
	Organics	Inorganics	Organics	Inorganics	Radionuclides/ radiation	Organics	Inorganics	Radionuclides/ radiation
Corytophanidae (casquehead lizards)								
Eublepharidae (eyelid geckos)			1 (1)			2 (1)		
Gekkonidae (geckos)		1 (1)		1 (1)			1 (1)	1 (1)
Pygopodidae (legless lizards)								
Dibamidae (blind lizards)								
Teiidae (whiptails and tegus)	5 (4)	1 (1)			1 (1)	2 (2)		7 (2)
Gymnophthalmidae (spectacled lizards)								
Lacertidae (wall lizards)		1 (1)	2 (2)	1 (1)		2 (2)		
Xantusiidae (night lizards)								
Scincidae (skinks)	3 (3)	2 (5)	3 (3)		1 (1)	2 (2)	1 (4)	
Cordylidae (spinytail lizards)								
Gerrhosauridae (plated and girdled lizards)								
Anguidae (glass lizards, alligator lizards, and lateral fold lizards)	1 (1)							
Xenosauridae (knob-scaled lizards)								
Varanidae (monitor lizards)	1 (2)	1 (8)	1 (2)			1 (2)		
Lanthanotidae (earless monitor lizards)								
Heleodermatidae (Gila monsters)		1 (2)						

Table 2. Residue or Bioaccumulation Data for Organic Contaminants.

Contaminant	Species	Tissue analyzed	Mean concentration, range	Units	Year	Location	Reference
Aldrin	*Pogona barbata*, bearded dragon	Intraperitoneal fat	<0.01	ppm	1970–1971	Undeveloped arid zone influenced by stock grazing, Northern Territory, Australia	Best (1973)
Aldrin	*Varanus giganteus*, perentie	Intraperitoneal fat	<0.01	ppm	1970–1971	Undeveloped arid zone influenced by stock grazing, Northern Territory, Australia	Best (1973)
Aldrin	*Varanus gouldii*, sand monitor	Intraperitoneal fat	<0.01	ppm	1970–1971	Developed tropical zone within a 90-km radius of Darwin, Northern Territory, Australia	Best (1973)
DDD or TDE	*Pogona barbata*, bearded dragon	Intraperitoneal fat	<0.01	ppm	1970–1971	Undeveloped arid zone influenced by stock grazing, Northern Territory, Australia	Best (1973)

Table 2. (Continued).

Contaminant	Species	Tissue analyzed	Mean concentration, range	Units	Year	Location	Reference
DDD or TDE	*Varanus giganteus*, perentie	Intraperitoneal fat	<0.01	ppm	1970–1971	Undeveloped arid zone influenced by stock grazing, Northern Territory, Australia	Best (1973)
DDD or TDE	*Varanus gouldii*, sand monitor	Intraperitoneal fat	<0.01	ppm	1970–1971	Developed tropical zone within a 90-km radius of Darwin, Northern Territory, Australia	Best (1973)
DDE	*Anolis carolinensis*, green anole	Carcass (whole body minus skin, head, feet, tail, and gastrointestinal tract)	0.12[a], <0.1–0.14	ppm, wet weight	1988	Citrus orchard, Haines City, Polk County, Florida, USA	Clark et al. (1995)
DDE	*Cnemidophorus gularis*, Texas spotted whiptail	Carcass (whole body minus skin, head, feet, tail, and gastrointestinal tract)	0.443[a], 0.14–0.96	ppm, wet weight	1988	Citrus orchards in Cameron and Hidalgo Counties, Lower Rio Grande Valley, Texas, USA	Clark et al. (1995)
DDE	*Cnemidophorus sexlineatus*, six-lined racerunner	Carcass (whole body minus skin, head, feet, tail, and gastrointestinal tract)	1.00[a], <0.1–9.6	ppm, wet weight	1988	Citrus orchard, Haines City, Polk County, Florida, USA	Clark et al. (1995)
DDE	*Cnemidophorus gularis*, Texas spotted whiptail	Whole body (minus intestinal tract)	3.5[a], 0.37–22	ppm, wet weight	1983	Pecos River drainage. Balmorhea, Texas, USA	White & Krynitsky (1986)

Table 2. (Continued).

Contaminant	Species	Tissue analyzed	Mean concentration, range	Units	Year	Location	Reference
DDE	*Cnemidophorus inornatus*, little striped whiptail	Whole body (minus intestinal tract)	<0.1[a]	ppm, wet weight	1983	Rio Grande River drainage (control site), Big Bend National Park, Texas, USA	White & Krynitsky (1986)
DDE	*Cnemidophorus inornatus*, little striped whiptail	Whole body (minus intestinal tract)	3.3[a], 1.0–10	ppm, wet weight	1983	Pecos and Rio Grande River drainages, Van Horn, Texas, USA	White & Krynitsky (1986)
DDE	*Cnemidophorus inornatus*, little striped whiptail	Whole body (minus intestinal tract)	0.41[a], 0.17–0.95	ppm, wet weight	1983	Rio Grande River drainage, Presidio, Texas, USA	White & Krynitsky (1986)
DDE	*Cnemidophorus inornatus*, little striped whiptail	Whole body (minus intestinal tract)	2.3[a], 0.1–16	ppm, wet weight	1983	Pecos and Rio Grande River drainages, Dell City, Texas, USA	White & Krynitsky (1986)
DDE	*Cnemidophorus inornatus*, little striped whiptail	Whole body (minus intestinal tract)	1.5[a], 0.64–4.8	ppm, wet weight	1983	Rio Grande River drainage, Fabens, Texas, USA	White & Krynitsky (1986)
DDE	*Cnemidophorus inornatus*, little striped whiptail	Whole body (minus intestinal tract)	17.2[a], 5.0–104	ppm, wet weight	1983	Pecos River drainage, Pecos, Texas, USA	White & Krynitsky (1986)
DDE	*Cnemidophorus inornatus*, little striped whiptail	Whole body (minus intestinal tract)	0.95^2, 0.18–10	ppm, wet weight	1983	Pecos River drainage, Artesia, New Mexico, USA	White & Krynitsky (1986)

Table 2. (Continued).

Contaminant	Species	Tissue analyzed	Mean concentration, range	Units	Year	Location	Reference
DDE	*Cnemidophorus inornatus*, little striped whiptail	Whole body (minus intestinal tract)	0.43[a], <0.1–1.5	ppm, wet weight	1983	Pecos River drainage, Roswell, New Mexico, USA	White & Krynitsky (1986)
DDE	*Cnemidophorus inornatus*, little striped whiptail	Whole body (minus intestinal tract)	0.36[a], 0.10–1.4	ppm, wet weight	1983	Pecos River drainage, Carlsbad, New Mexico, USA	White & Krynitsky (1986)
DDE	*Cnemidophorus inornatus*, little striped whiptail and *Cnemidophorus tesselatus*, checkered whiptail	Whole body (minus intestinal tract)	<0.1[a]	ppm, wet weight	1983	Pecos River drainage (control site), Fort Sumner, New Mexico, USA	White and Krynitsky (1986)
DDE	*Cnemidophorus gularis*, Texas spotted whiptail; *Cnemidophorus inornatus*, little striped whiptail; *Cnemidophorus tesselatus*, checkered whiptail	While body (minus intestinal tract)	0.47[a], <0.1–4.8	ppm, wet weight	1983	Rio Grande River drainage, USA	White & Krynitsky (1986)

Table 2. (Continued).

Contaminant	Species	Tissue analyzed	Mean concentration, range	Units	Year	Location	Reference
DDE	*Cnemidophorus gularis*, Texas spotted whiptail; *Cnemidophorus inornatus*, little striped whiptail; *Cnemidophorus tesselatus*, checkered whiptail	Whole body (minus intestinal tract)	1.8[a], <0.1–104	ppm, wet weight	1983	Pecos River drainage, USA	White & Krynitsky (1986)
DDE	*Morethia boulengeri*, Boulenger's morethia	Body fat	0.1	ppm, wet weight	1972	Saddleworth-Riverton Area, South Australia	Birks & Olsen (1987)
DDE	*Pogona barbata*, bearded dragon	Intraperitoneal fat	<0.01	ppm	1970–1971	Undeveloped arid zone influenced by stock grazing, Northern Territory, Australia	Best (1973)
DDE	*Varanus giganteus*, perentie	Intraperitoneal fat	<0.01	ppm	1970–1971	Undeveloped arid zone influenced by stock grazing, Northern Territory, Australia	Best (1973)

Table 2. (Continued).

Contaminant	Species	Tissue analyzed	Mean concentration, range	Units	Year	Location	Reference
DDE	*Varanus gouldii*, sand monitor	Intraperitoneal fat	0.185, 0.04–0.33	ppm	1970–1971	Developed tropical zone within a 90-km radius of Darwin, Northern Territory, Australia	Best (1973)
DDE	*Cnemidophorus tesselatus*, checkered whiptail	Whole body	June: 4.2[b] July: 2.7[b] August: 1.2[b] September: 0.5[b]	ppm	1967	Presidio Basin, Rio Grande Valley, Southwest Texas, USA	Saxon (1970)
DDT	*Pogona barbata*, bearded dragon	Intraperitoneal fat	<0.01	ppm	1970–1971	Undeveloped arid zone influenced by stock grazing, Northern Territory, Australia	Best (1973)
DDT	*Varanus giganteus*, perentie	Intraperitoneal fat	<0.01	ppm	1970–1971	Undeveloped arid zone influenced by stock grazing, Northern Territory, Australia	Best (1973)
DDT	*Varanus gouldii*, sand monitor	Intraperitoneal fat	0.02, <0.01–0.03	ppm	1970–1971	Developed tropical zone within a 90-km radius of Darwin, Northern Territory, Australia	Best (1973)

Table 2. (Continued).

Contaminant	Species	Tissue analyzed	Mean concentration, range	Units	Year	Location	Reference
DDT	*Cnemidophorus tesselatus*, checkered whiptail	Whole body	June: 0.2[b] July: 0.6[b] August: 0.5[b] September: 0[b]	ppm	1967	Presidio Basin, Rio Grande Valley, Southwest Texas, USA	Saxon (1970)
Dicofol	*Cnemidophorus gularis*, Texas spotted whiptail	Carcass (whole body minus skin, head, feet, tail, and gastrointestinal tract)	0.712[a], <0.1–12	ppm, wet weight	1988	Citrus orchards in Cameron and Hidalgo Counties, Lower Rio Grande Valley, Texas, USA (Dicofol applied 2.8 kg active ingredient/ha 11 months before lizards were collected)	Clark et al. (1995)
Dicofol	*Cnemidophorus sextlineatus*, six-lined racerunner	Carcass (whole body minus skin, head, feet, tail, and gastrointestinal tract)	0.15[a], <0.1–0.20	ppm, wet weight	1988	Citrus orchard, Haines City, Polk County, Florida, USA (Dicofol applied 1.7–2.0 kg active ingredient/ha monthly 2 months before lizards were collected)	Clark et al. (1995)

Table 2. (Continued).

Contaminant	Species	Tissue analyzed	Mean concentration, range	Units	Year	Location	Reference
Dieldrin	*Pogona barbata*, bearded dragon	Intraperitoneal fat	<0.01	ppm	1970–1971	Undeveloped arid zone influenced by stock grazing, Northern Territory, Australia	Best (1973)
Dieldrin	*Varanus giganteus*, perentie	Intraperitoneal fat	0.03	ppm	1970–1971	Undeveloped arid zone influenced by stock grazing, Northern Territory, Australia	Best (1973)
Dieldrin	*Varanus gouldii*, sand monitor	Intraperitoneal fat	0.02, <0.01–0.03	ppm	1970–1971	Developed tropical zone within a 90-km radius of Darwin, Northern Territory, Australia	Best (1973)
Endrin	*Pogona barbata*, bearded dragon	Intraperitoneal fat	<0.01	ppm	1970–1971	Undeveloped arid zone influenced by stock grazing, Northern Territory, Australia	Best (1973)
Endrin	*Varanus giganteus*, perentie	Intraperitoneal fat	<0.01	ppm	1970–1971	Undeveloped arid zone influenced by stock grazing, Northern Territory, Australia	Best (1973)

Table 2. (Continued).

Contaminant	Species	Tissue analyzed	Mean concentration, range	Units	Year	Location	Reference
Endrin	*Varanus gouldii*, sand monitor	Intraperitoneal fat	<0.01	ppm	1970–1971	Developed tropical zone within a 90-km radius of Darwin, Northern Territory, Australia	Best (1973)
Hexachlorobenzene (HCB)	*Pogona barbata*, bearded dragon	Intraperitoneal fat	<0.01	ppm	1970–1971	Undeveloped arid zone influenced by stock grazing, Northern Territory, Australia	Best (1973)
HCB	*Varanus giganteus*, perentie	Intraperitoneal fat	<0.01	ppm	1970–1971	Undeveloped arid zone influenced by stock grazing, Northern Territory, Australia	Best (1973)
HCB	*Varanus gouldii*, sand monitor	Intraperitoneal fat	0.015, <0.01–0.02	ppm	1970–1971	Developed tropical zone within a 90-km radius of Darwin, Northern Territory, Australia	Best (1973)
Lindane	*Pogona barbata*, bearded dragon	Intraperitoneal fat	<0.01	ppm	1970–1971	Undeveloped arid zone influenced by stock grazing, Northern Territory, Australia	Best (1973)

Table 2. (Continued).

Contaminant	Species	Tissue analyzed	Mean concentration, range	Units	Year	Location	Reference
Lindane	*Varanus giganteus*, perentie	Intraperitoneal fat	<0.01	ppm	1970–1971	Undeveloped arid zone influenced by stock grazing, Northern Territory, Australia	Best (1973)
Lindane	*Varanus gouldii*, sand monitor	Intraperitoneal fat	0.065, <0.01–0.12	ppm	1970–1971	Developed tropical zone within a 90-km radius of Darwin, Northern Territory, Australia	Best (1973)
Methyl parathion	*Cnemidophorus tesselatus*, checkered whiptail	Whole body	June: 7.7[b] July: 8.1[b] August: 1.0[b] September: 0.1[b]	ppm	1967	Presidio Basin, Rio Grande Valley, Southwest Texas, USA	Saxon (1970)
Mirex	*Sceloporus undulatus*, eastern fence lizard	Whole body	1 mon post-treatment: 0.05 3 mon post-treatment: 0.35, 0.19–0.51 6 mon post-treatment: 0.15, 0.12–0.18 9 mon post-treatment: 0.08	ppm, fresh weight	1972–1974	Dee Dot Ranch, Duval and St. Johns Counties, near Jacksonville, Florida, USA (10-5 bait formulation of 0.1% mirex applied at 1.12 kg/ha)	Wheeler et al. (1977)

Table 2. (Continued).

Contaminant	Species	Tissue analyzed	Mean concentration, range	Units	Year	Location	Reference
Mirex	*Scincella lateralis*, ground skink	Whole body	1 yr post-treatment: 0.04 18 mon post-treatment: 0.30 2 yr post-treatment: <0.01 9 mon post-treatment: 0.11	ppm, fresh weight	1972–1974	Dee Dot Ranch, Duval and St. Johns Counties, near Jacksonville, Florida, USA (10-5 bait formulation of 0.1% mirex applied at 1.12 kg/ha)	Wheeler et al. (1977)
Mirex	*Cnemidophorus sexlineatus*, six-lined racerunner	Whole body	2 yr post-treatment: <0.01	ppm, fresh weight	1972–1974	Dee Dot Ranch, Duval and St. Johns Counties, near Jacksonville, Florida, USA (10-5 bait formulation of 0.1% mirex applied at 1.12 kg/ha)	Wheeler et al. (1977)

Table 2. (Continued).

Contaminant	Species	Tissue analyzed	Mean concentration, range	Units	Year	Location	Reference
Mirex	*Ophisaurus* sp. glass lizard	Whole body	9 mon post-treatment: <0.01	ppm, fresh weight	1972–1974	Dee Dot Ranch, Duval and St. Johns Counties, near Jacksonville, Florida, USA (10-5 bait formulation of 0.1% mirex applied at 1.12 kg/ha)	Wheeler et al. (1977)
Mirex	*Scincella lateralis*, ground skink	Whole body	3 mon post-treatment: 0.34; 6 mon post-treatment: 0.22; 1 yr post-treatment: <0.01	ppm	1971–1972	Tift County, Southwest Georgia, USA (1.12 g/ha application rate)	Wojcik et al. (1975)
Mirex	*Scincella lateralis*, ground skink	Whole body	6 mon post-treatment: 0.66	ppm	1971–1972	Worth County, Southwest Georgia, USA (4.20 g/ha application rate)	Wojcik et al. (1975)
Mirex	*Eumeces laticeps*, broadhead skink	Whole body	6 mon post-treatment: <0.01	ppm	1971–1972	Worth County, Southwest Georgia, USA (1.12 g/ha application rate)	Wojcik et al. (1975)
Mirex	*Cnemidophorus sexlineatus*, six-lined racerunner	Whole body	3 mon post-treatment: 0.93	ppm	1971–1972	Turner County, Southwest Georgia, USA (2.10 g/ha application rate)	Wojcik et al. (1975)

Table 2. (Continued).

Contaminant	Species	Tissue analyzed	Mean concentration, range	Units	Year	Location	Reference
Mirex	*Cnemidophorus sexlineatus*, six-lined racerunner	Whole body	1 mon post-treatment: 0.63 3 mon post-treatment: 0.07 1 yr post-treatment: 0.40	ppm	1971–1972	Turner County, Southwest Georgia, USA (4.20 g/ha application rate)	Wojcik et al. (1975)
Monodechlorinated dicofol (DCD)	*Cnemidophorus gularis*, Texas spotted whiptail	Carcass (whole body minus skin, head, feet, tail, and gastrointestinal tract)	0.868[a], <0.1–15	ppm, wet weight	1988	Citrus orchards in Cameron and Hidalgo Counties, Lower Rio Grande Valley, Texas, USA (Dicofol applied 2.8 kg active ingredient/ha 11 months before lizards were collected)	Clark et al. (1995)
Parathion	*Cnemidophorus tesselatus*, checkered whiptail	Whole body	June: 0[b] July: 0[b] August: 0.2[b] September: 0.1[b]	ppm	1967	Presidio Basin, Rio Grande Valley, Southwest Texas, USA	Saxon (1970)

[a]Geometric mean.
[b]Mean

able from six studies (Table 3). Contaminant concentrations were available as whole-body concentrations as well as those for specific organs. Whole-body contaminant concentrations are important if the lizard is being considered as a food source or prey item, whereas concentrations in specific organs can be used in studies on the lizard itself.

A. Organic Contaminants

Of the seven studies reviewed for lizard organic contaminant concentrations, data were available for seven families of lizards, which included 13 species (Table 1). All available lizard organic contaminant residue data were for pesticides (Table 2).

Pesticide concentrations in checkered whiptails were studied in the Presidio Basin, Rio Grande Valley in Southwest Texas, by Saxon (1970). Cotton is the major crop grown in the Presidio Basin, and pesticides are applied to cotton fields each season. Saxon (1970) analyzed whiptails collected in June, July, August, and September from sites within a 48.3-km radius of the Presidio Basin for whole-body concentrations of DDT (and metabolites DDD and DDE), methyl parathion, and parathion. Checkered whiptails contained the highest levels of insecticides in June and July, which coincided with the months of heavy pesticide application (Table 2). DDD was not detected in any of the whiptails during any month. Consumption of contaminated food was probably the most important pathway for the transfer of pesticides. Saxon's study (1970) added to the information on whiptail contamination in the Presidio Basin obtained earlier by Culley and Applegate (1967a,b) [reviewed by Hall (1980)]. They analyzed the tail muscle, brain tissue, liver tissue, stomach contents, postcoelomic fat, and eggs of whiptail lizards collected in the same areas of the Presidio Basin in June, July, and August for HCB, DDT (and metabolites DDD and DDE), endrin, malathion, methyl parathion, and parathion.

Little striped, checkered, and Texas spotted whiptails in New Mexico and Texas were analyzed for whole-body (minus intestinal tract) concentrations of DDE by White and Krynitsky (1986). Concentrations of DDE were significantly higher (up to 104 ppm, wet weight) in whiptails from the Pecos River drainage than those taken from the Rio Grande drainage (Table 2). Thus, even though DDT usage was banned in the U.S. in 1972, lizards collected in 1983 contained high concentrations of DDE, which was acquired while foraging in or near agricultural fields (White and Krynitsky 1986).

Residues of DDT (and metabolites DDD and DDE) and dicofol (and metabolite monodechlorinated dicofol), which is structurally similar to DDT, were analyzed in carcasses (whole body minus skin, head, feet, tail, and gastrointestinal tract) of lizards collected from Florida, Texas, and California (Clark et al. 1995). Texas spotted whiptails collected from the Lower Rio Grande Valley, Texas, contained the highest concentrations of dicofol, but also contained DDE (Table 2). The highest DDE concentrations were found in six-lined racerunners collected from a citrus orchard in Florida 16 yr after DDT was banned; one race-

Table 3. Residue or Bioaccumulation Data for Inorganic Contaminants.

Contaminant	Species	Tissue analyzed	Mean concentration, range	Units	Year	Location	Reference
Aluminum	*Agama stellio stellio*, starred agama	Whole body (minus liver and digestive tract): Liver tissue:	769.49, 11.04–1,527.94 119.98, 48.60–191.36	ppm, dry weight	1996	T Area (edge and 4–5 km E of the center of Thessaloniki, Greece, altitude of 500 m)	Loumbourdis (1997)
Aluminum	*Agama stellio stellio*, starred agama	Whole body (minus liver and digestive tract): Liver tissue:	1,601.70, 1,038.10–2,165.30 132.99, 61.89–204.09	ppm, dry weight	1996	K Area (near village of Kolhiko, 25 km NE of Thessaloniki, Greece, altitude of 50 m, near cultivated areas)	Loumbourdis (1997)
Barium	*Agama stellio stellio*, starred agama	Whole body (minus liver and digestive tract): Liver tissue:	49.76, 29.97–69.52 13.04, 5.74–20.34	ppm, dry weight	1996	T Area (edge and 4–5 km E of the center of Thessaloniki, Greece, altitude of 500 m)	Loumbourdis (1997)
Barium	*Agama stellio stellio*, starred agama	Whole body (minus liver and digestive tract): Liver tissue:	122.88, 74.83–170.93 22.77, 11.64–33.90	ppm, dry weight	1996	K Area (near village of Kolhiko, 25 km NE of Thessaloniki, Greece, altitude of 50 m, near cultivated areas)	Loumbourdis (1997)

Table 3. (Continued).

Contaminant	Species	Tissue analyzed	Mean concentration, range	Units	Year	Location	Reference
Cadmium	*Agama stellio stellio*, starred agama	Whole body (minus liver and digestive tract): Liver tissue:	1.01, 0.47–1.55 9.87, 5.16–14.58	ppm, dry weight	1996	T Area (edge and 4–5 km E of the center of Thessaloniki, Greece, altitude of 500 m)	Loumbourdis (1997)
Cadmium	*Agama stellio stellio*, starred agama	Whole body (minus liver and digestive tract): Liver tissue:	1.36, 0.75–1.97 27.18, 14.59–39.77	ppm, dry weight	1996	K Area (near village of Kolhiko, 25 km NE of Thessaloniki, Greece, altitude of 50 m, near cultivated areas)	Loumbourdis (1997)
Cadmium	*Hemidactylus mabouia*, cosmopolitan house gecko	Whole body (minus stomach contents): Liver tissue:	45–82 (range of means), 11–223 402–1,299 (range of means), 132–4,007	ppb, dry weight	1982	Nine sites in Porto Alegre, Brazil	Schmidt (1984)
Cesium	*Agama stellio stellio*, starred agama	Whole body (minus liver and digestive tract): Liver tissue:	0.26, 0.06–0.46 0.48, 0.14–0.82	ppm, dry weight	1996	T Area (edge and 4–5 km E of the center of Thessaloniki, Greece, altitude of 500 m)	Loumbourdis (1997)
Cesium	*Agama stellio stellio*, starred agama	Whole body (minus liver and digestive tract): Liver tissue:	0.16, 0.151–0.169 0.98, 0–2.15	ppm, dry weight	1996	K Area (near village of Kolhiko, 25 km NE of Thessaloniki, Greece, altitude of 50 m, near cultivated areas)	Loumbourdis (1997)

Table 3. (Continued).

Contaminant	Species	Tissue analyzed	Mean concentration, range	Units	Year	Location	Reference
Chromium	*Agama stellio stellio*, starred agama	Whole body (minus liver and digestive tract):	4.16, 2.71–5.61	ppm, dry weight	1996	T Area (edge and 4–5 km E of the center of Thessaloniki, Greece, altitude of 500 m)	Loumbourdis (1997)
		Liver tissue:	2.37, 1.78–2.96				
Chromium	*Agama stellio stellio*, starred agama	Whole body (minus liver and digestive tract):	5.57, 4.30–6.84	ppm, dry weight	1996	K Area (near village of Kolhiko, 25 km NE of Thessaloniki, Greece, altitude of 50 m, near cultivated areas)	Loumbourdis (1997)
		Liver tissue:	1.35, 0.58–2.14				
Cobalt	*Agama stellio stellio*, starred agama	Whole body (minus liver and digestive tract):	2.53, 1.90–3.16	ppm, dry weight	1996	T Area (edge and 4–5 km E of the center of Thessaloniki, Greece, altitude of 500 m)	Loumbourdis (1997)
		Liver tissue:	3.50, 1.11–5.89				
Cobalt	*Agama stellio stellio*, starred agama	Whole body (minus liver and digestive tract):	3.62, 3.00–4.24	ppm, dry weight	1996	K Area (near village of Kolhiko, 25 km NE of Thessaloniki, Greece, altitude of 50 m, near cultivated areas)	Loumbourdis (1997)
		Liver tissue:	5.08, 3.64–6.52				

Table 3. (Continued).

Contaminant	Species	Tissue analyzed	Mean concentration, range	Units	Year	Location	Reference
Copper	*Agama stellio stellio*, starred agama	Whole body (minus liver and digestive tract): Liver tissue:	27.44, 21.38–33.50 139.67, 101.96–177.38	ppm, dry weight	1996	T Area (edge and 4–5 km E of the center of Thessaloniki, Greece, altitude of 500 m)	Loumbourdis (1997)
Copper	*Agama stellio stellio*, starred agama	Whole body (minus liver and digestive tract): Liver tissue:	27.14, 22.84–31.44 209.09, 86.86–331.32	ppm, dry weight	1996	K Area (near village of Kolhiko, 25 km NE of Thessaloniki, Greece, altitude of 50 m, near cultivated areas)	Loumbourdis (1997)
Copper	*Egernia napoleonsis*, Napoleon skink	Liver tissue	10.0	ppm, dry weight	1953	Cheyne Beach, Australia	Beck (1956)
Lead	*Agama stellio stellio*, starred agama	Whole body (minus liver and digestive tract): Liver tissue:	12.81, 9.05–16.57 6.79, 3.97–9.61	ppm, dry weight	1996	T Area (edge and 4–5 km E of the center of Thessaloniki, Greece, altitude of 500 m)	Loumbourdis (1997)
Lead	*Agama stellio stellio*, starred agama	Whole body (minus liver and digestive tract): Liver tissue:	15.65, 10.14–21.16 13.32, 8.29–18.35	ppm, dry weight	1996	K Area (near village of Kolhiko, 25 km NE of Thessaloniki, Greece, altitude of 50 m, near cultivated areas)	Loumbourdis (1997)
Lead	*Podarcis muralis*, common wall lizard	Scales	7.00, 4.7–9.3	ppm, dry weight	Late 1980s	Rural sites in Punjab, India	Kaur (1988)

Table 3. (Continued).

Contaminant	Species	Tissue analyzed	Mean concentration, range	Units	Year	Location	Reference
Lead	*Podarcis muralis*, common wall lizard	Scales	150.53, 140.53–160.53	ppm, dry weight	Late 1980s	Urban sites in Punjab, India	Kaur (1988)
Lead	*Hemidactylus mabouia*, cosmopolitan house gecko	Whole body (minus stomach contents):	1.8–13.0 (range of means), 0.5–30.3	ppm, dry weight	1982	Nine sites in Porto Alegre, Brazil	Schmidt (1984)
		Liver tissue:	2.0–5.1 (range of means), 1.3–12.4				
Manganese	*Agama stellio stellio*, starred agama	Whole body (minus liver and digestive tract):	40.71, 15.90–65.52	ppm, dry weight	1996	T Area (edge and 4–5 km E of the center of Thessaloniki, Greece, altitude of 500 m)	Loumbourdis (1997)
		Liver tissue:	41.02, 12.66–69.38				
Manganese	*Agama stellio stellio*, starred agama	Whole body (minus liver and digestive tract):	61.09, 55.15–67.03	ppm, dry weight	1996	K Area (near village of Kolhiko, 25 km NE of Thessaloniki, Greece, altitude of 50 m, near cultivated areas)	Loumbourdis (1997)
		Liver tissue:	2.03, 22.32–81.74				

Table 3. (Continued).

Contaminant	Species	Tissue analyzed	Mean concentration, range	Units	Year	Location	Reference
Mercury	*Ameiva exsul*, Puerto Rican ameiva	Whole body	<80	ppb, wet weight	1988	Three estuaries in Puerto Rico (Humacao Marshes and Roosevelt Roads on E coast and Boqueron on SW coast	Burger et al. (1992)
Molybdenum	*Agama stellio stellio*, starred agama	Whole body (minus liver and digestive tract):	1.24, 0.81–1.67	ppm, dry weight	1996	T Area (edge and 4–5 km E of the center of Thessaloniki, Greece, altitude of 500 m)	Loumbourdis (1997)
		Liver tissue:	8.32, 5.70–10.94				
Molybdenum	*Agama stellio stellio*, starred agama	Whole body (minus liver and digestive tract):	1.39, 0.56–2.22	ppm, dry weight	1996	K Area (near village of Kolhiko, 25 km NE of Thessaloniki, Greece, altitude of 50 m, near cultivated areas)	Loumbourdis (1997)
		Liver tissue:	7.51, 5.09–9.93				
Nickel	*Agama stellio stellio*, starred agama	Whole body (minus liver and digestive tract):	33.83, 11.71–55.95	ppm, dry weight	1996	T Area (edge and 4–5 km E of the center of Thessaloniki, Greece, altitude of 500 m)	Loumbourdis (1997)
		Liver tissue:	3.60, 0.49–6.71				
Nickel	*Agama stellio stellio*, starred agama	Whole body (minus liver and digestive tract):	47.52, 88.60–162.39	ppm, dry weight	1996	K Area (near village of Kolhiko, 25 km NE of Thessaloniki, Greece, altitude of 50 m, near cultivated areas)	Loumbourdis (1997)
		Liver tissue:	7.33, 2.26–12.40				

Table 3. (Continued).

Contaminant	Species	Tissue analyzed	Mean concentration, range	Units	Year	Location	Reference
Rubidium	*Agama stellio stellio*, starred agama	Whole body (minus liver and digestive tract):	30.96, 25.58–36.34	ppm, dry weight	1996	T Area (edge and 4–5 km E of the center of Thessaloniki, Greece, altitude of 500 m)	Loumbourdis (1997)
		Liver tissue:	33.92, 25.29–42.55				
Rubidium	*Agama stellio stellio*, starred agama	Whole body (minus liver and digestive tract):	33.73, 20.92–46.54	ppm, dry weight	1996	K Area (near village of Kolhiko, 25 km NE of Thessaloniki, Greece, altitude of 50 m, near cultivated areas)	Loumbourdis (1997)
		Liver tissue:	35.05, 19.24–50.86				
Strontium	*Agama stellio stellio*, starred agama	Whole body (minus liver and digestive tract):	163.23, 124.61–254.81	ppm, dry weight	1996	T Area (edge and 4–5 km E of the center of Thessaloniki, Greece, altitude of 500 m)	Loumbourdis (1997)
		Liver tissue:	16.48, 4.62–28.34				
Strontium	*Agama stellio stellio*, starred agama	Whole body (minus liver and digestive tract):	396.12, 333.01–459.23	ppm, dry weight	1996	K Area (near village of Kolhiko, 25 km NE of Thessaloniki, Greece, altitude of 50 m, near cultivated areas)	Loumbourdis (1997)
		Liver tissue:	40.24, 27.52–52.96				

Table 3. (Continued).

Contaminant	Species	Tissue analyzed	Mean concentration, range	Units	Year	Location	Reference
Zinc	*Agama stellio stellio*, starred agama	Whole body (minus liver and digestive tract): Liver tissue:	608.99, 502.35–715.63 614.55, 449.29–779.81	ppm, dry weight	1996	T Area (edge and 4–5 km E of the center of Thessaloniki, Greece, altitude of 500 m)	Loumbourdis (1997)
Zinc	*Agama stellio stellio*, starred agama	Whole body (minus liver and digestive tract): Liver tissue:	643.47, 591.44–695.50 794.14, 616.11–972.17	ppm, dry weight	1996	K Area (near village of Kolhiko, 25 km NE of Thessaloniki, Greece, altitude of 50 m, near cultivated areas)	Loumbourdis (1997)
Zinc	*Iguana iguana*, green iguana	Plasma	2.30, 1.1–3.3	µg/mL	10-yr period	Louisiana State University Museum of Natural History, California Academy of Sciences, San Diego Zoo, and wild caught	Lance et al. (1995)
Zinc	*Cyclura* sp., rhinoceros iguana	Plasma	4.1	µg/mL	10-yr period	Louisiana State University Museum of Natural History, California Academy of Sciences, San Diego Zoo, and wild caught	Lance et al. (1995)

Table 3. (Continued).

Contaminant	Species	Tissue analyzed	Mean concentration, range	Units	Year	Location	Reference
Zinc	*Corucia zebrata*, Solomon Island skink	Plasma	6.6, 4.9–7.6	µg/mL	10-yr period	Louisiana State University Museum of Natural History, California Academy of Sciences, San Diego Zoo, and wild caught	Lance et al. (1995)
Zinc	*Tiliqua scincoides*, eastern bluetongue skink	Plasma	8.5	µg/mL	10-yr period	Louisiana State University Museum of Natural History, California Academy of Sciences, San Diego Zoo, and wild caught	Lance et al. (1995)
Zinc	*Tiliqua rugosa*, shingleback skink	Plasma	5.0	µg/mL	10-yr period	Louisiana State University Museum of Natural History, California Academy of Sciences, San Diego Zoo, and wild caught	Lance et al. (1995)
Zinc	*Tiliqua nigrolutea*, blotched bluetongue skink	Plasma	9.8	µg/mL	10-yr period	Louisiana State University Museum of Natural History, California Academy of Sciences, San Diego Zoo, and wild caught	Lance et al. (1995)

Table 3. (Continued).

Contaminant	Species	Tissue analyzed	Mean concentration, range	Units	Year	Location	Reference
Zinc	*Heloderma horridum*, beaded lizard	Plasma	0.20, 0.02–0.57	µg/mL	10-yr period	Louisiana State University Museum of Natural History, California Academy of Sciences, San Diego Zoo, and wild caught	Lance et al. (1995)
Zinc	*Heloderma suspectum*, Gila monster	Plasma	0.65, 0.21–1.62	µg/mL	10-yr period	Louisiana State University Museum of Natural History, California Academy of Sciences, San Diego Zoo, and wild caught	Lance et al. (1995)
Zinc	*Varanus salvadorii*, crocodile monitor	Plasma	3.05	µg/mL	10-yr period	Louisiana State University Museum of Natural History, California Academy of Sciences, San Diego Zoo, and wild caught	Lance et al. (1995)
Zinc	*Varanus acanthurus*, ridgetail monitor	Plasma	7.02, 6.48–7.57	µg/mL	10-yr period	Louisiana State University Museum of Natural History, California Academy of Sciences, San Diego Zoo, and wild caught	Lance et al. (1995)

Table 3. (Continued).

Contaminant	Species	Tissue analyzed	Mean concentration, range	Units	Year	Location	Reference
Zinc	*Varanus exanthematicus*, savannah monitor	Plasma	2.38, 1.97–2.68	µg/mL	10-yr period	Louisiana State University Museum of Natural History, California Academy of Sciences, San Diego Zoo, and wild caught	Lance et al. (1995)
Zinc	*Varanus grayi*, Gray's monitor	Plasma	9.59, 0.31–24.38	µg/mL	10-yr period	Louisiana State University Museum of Natural History, California Academy of Sciences, San Diego Zoo, and wild caught	Lance et al. (1995)
Zinc	*Varanus prasinus*, emerald monitor	Plasma	1.54	µg/mL	10-yr period	Louisiana State University Museum of Natural History, California Academy of Sciences, San Diego Zoo, and wild caught	Lance et al. (1995)

Table 3. (Continued).

Contaminant	Species	Tissue analyzed	Mean concentration, range	Units	Year	Location	Reference
Zinc	*Varanus albigularis*, white-throated monitor	Plasma	5.35, 2.32–6.79	µg/mL	10-yr period	Louisiana State University Museum of Natural History, California Academy of Sciences, San Diego Zoo, and wild caught	Lance et al. (1995)
Zinc	*Varanus indicus*, mangrove monitor	Plasma	2.89, 2.15–3.63	µg/mL	10-yr period	Louisiana State University Museum of Natural History, California Academy of Sciences, San Diego Zoo, and wild caught	Lance et al. (1995)
Zinc	*Varanus komodoensis*, Komodo dragon	Plasma	2.73, 0.61–5.54	µg/mL	10-yr period	Louisiana State University Museum of Natural History, California Academy of Sciences, San Diego Zoo, and wild caught	Lance et al. (1995)
Zinc	*Hemidactylus mabouia*, cosmopolitan house gecko	Whole body (minus stomach contents): Liver tissue:	106–128 (range of means), 27–303 125–226 (range of means), 61–544	ppm, dry weight	1982	Nine sites in Porto Alegre, Brazil	Schmidt (1984)

runner contained dicofol. Green anoles from the Florida citrus orchard contained small amounts of DDE and no dicofol. Side-blotched and western fence lizards from sites near California cotton fields (Kings County) contained no residues of either dicofol or DDT.

Mirex residues were monitored in eastern fence lizards, ground skinks, six-lined racerunners, and glass lizards for 2 yr after treatment of a 8,000-ha ranch in Florida with 0.1% mirex applied at 1.12 kg/ha to control fire ants (*Solenopsis saevissima*) (Wheeler et al. 1977). In general, low levels were found in lizards throughout the 24-mon period (Table 2). Ground skinks, broadhead skinks, and six-lined racerunners were analyzed for mirex up to 1 yr after treatment of three large areas in southwest Georgia for fire ant control using four rates of mirex application (Wojcik et al. 1975). Mirex residues were found in lizards during the entire 1-yr period and declined over time (Table 2).

The intraperitoneal fat of bearded dragons, perenties, and sand monitors collected from undeveloped and developed arid and tropical zones of the Northern Territory, Australia, was analyzed for DDT, DDE, DDD, dieldrin, lindane, HCB, endrin, and aldrin (Best 1973). For the most part, concentrations of all pesticide residues were below the limits of detection in perenties and bearded dragons from the undeveloped arid zone area (Table 2). DDD, endrin, and aldrin residues were below the detection limits in intraperitoneal fat in sand monitors collected from the developed tropical zone area. However, low levels of DDT, DDE, dieldrin, lindane, and HCB were found in sand monitors (Table 2). Birks and Olsen (1987) analyzed the body fat of lizards for DDT plus metabolites, HCB, lindane, aldrin, and dieldrin collected from the Saddleworth-Riverton area in South Australia. The body fat of Boulenger's morethia (a skink) collected from roadside vegetation contained 0.1 mg/kg (wet weight) DDE, but other pesticides were not detected in these lizards or in stomach fat from bearded dragons (Birks and Olsen 1987) (Table 2).

B. Inorganic Contaminants

Very few studies were found in which inorganic or heavy metal concentrations in lizards were analyzed (Table 3). Information was available for 21 species of lizards, representing eight families (Table 1). Two studies were included in this review because the data represented natural or background levels contained in lizards collected from uncontaminated sites. The oldest study included copper concentrations found in lizard livers (Beck 1956). Lance et al. (1995) analyzed blood plasma in 16 different species of lizards to determine normal levels of zinc that would be present in lizard blood plasma. In a study of mercury levels in animals from different trophic levels in three estuaries in Puerto Rico, mercury was not detected in Puerto Rican ameivas (Burger et al. 1992).

Loumbourdis (1997) analyzed aluminum, barium, cadmium, cesium, chromium, cobalt, copper, lead, manganese, molybdenum, nickel, rubidium, strontium, and zinc concentrations in the whole body (minus liver and digestive tract) and liver of the starred agama in an urban, high-altitude site and an agricultural, low-altitude area in Greece. This study was the first in which so many heavy metals were analyzed in lizards. The main purpose of the study was to determine whether lizards could be used as bioindicators of heavy metal pollution. Overall, metal concentrations were highest in lizards from the low-altitude, agricultural area (Table 3). A possible explanation for this is that the urban site lies at a higher altitude, and a high percentage of pollutants tend to accumulate at lower altitudes. Moreover, the direction of the prevailing winds is such that heavy metals are pushed to the low-altitude area. Another explanation could be the use of pesticides and fertilizers that contain metals. The results of the study indicated that lizards accumulate heavy metals and could be used as bioindicators of pollution.

The scales of common wall lizards living in urban and rural sites in Punjab, India were analyzed for lead concentrations (Kaur 1988). Scales of wall lizards from urban areas contained significantly higher amounts of lead than those from rural sites (Table 3). The results indicated that high lead exposure to reptiles living near busy roads in urban areas could be detrimental to the health and longevity of lizards (Kaur 1988).

Schmidt (1984) conducted a study using the cosmopolitan house gecko to detect heavy metal contamination in Porto Alegre, Brazil. Whole-body concentrations (minus stomach contents) of lead, cadmium, and zinc were analyzed in geckos collected from nine locations throughout the city (Table 3). Lead concentrations were related to traffic levels at a particular location; the higher the traffic, the higher the lead concentration. Liver tissue was analyzed from some geckos (Table 3). Cadmium, lead, and zinc accumulated in the liver. Feces analyzed from geckos collected from the Tres Figueiras site contained much higher lead (10 fold) and cadmium (20 fold) concentrations than did whole bodies, indicating active excretion of heavy metals. Overall, the results of the study indicated that geckos accumulated heavy metals and were a good indicator of heavy metal pollution.

Schmidt (1981) conducted an earlier study during the late 1970s in Saarbrücken, Germany, to determine if lizards were a good indicator of heavy metal pollution. Lizards (common Eurasian lizards, *Lacerta muralis*; viviparous lizards, *Lacerta vivipara*; and slow worms, *Anguis fragilis*) were collected from various locations within and outside of the city of Saarbrücken and analyzed for whole-body concentrations of lead and cadmium. Quantitative data were not available for inclusion in this review. However, the results of the study indicated that heavy metal concentrations decreased with increasing distance from urban locations and that lizards are useful bioindicators of heavy metal contamination in urban areas (Schmidt 1981).

V. Lethal and Sublethal Effects Data

Studies concerning the lethal and sublethal effects of contaminants were reviewed for 13 families of lizards (Table 1). Eleven studies were reviewed for lizard effects data for organic contaminants (Table 4). Seven studies included

Table 4. Lethal and Sublethal Effects Data for Organic Contaminants.

Contaminant	Species	Exposure	Year	Location	Effects	Reference
Azinphos-methyl	*Anolis carolinensis*, green anole	98% dissolved in corn oil, dose administered in laboratory and equivalent to 5% body weight of each lizard	1980	Captured from Merritt Island National Wildlife Refuge, Florida and in Gainesville, Florida, USA	LD_{50} = 98 mg/kg Levels of cholinesterase activity (inhibition) as a % of normal related to 50% mortality: 45.5%	Hall & Clark (1982)
Benzomate	*Eumeces marginatus*, Ousima skink	1% and 2% emulsifiable concentration	1977	Laboratory	1% concentration: no effect 2% concentration: death; time to death range 1–3 hr	Kihara & Yamashita (1978)
Carbaryl	*Eumeces marginatus*, Ousima skink	1% and 2% wettable powder and 3% dust	1977	Laboratory	No effect at any of the three doses	Kihara & Yamashita (1978)
Chlorfenvinphos	*Eumeces marginatus*, Ousima skink	1% and 2% emulsifiable concentration	1977	Laboratory	Both concentrations: death; time to death 1 hr	Kihara & Yamashita (1978)
Chlorpyrifos	*Eumeces marginatus*, Ousima skink	1% and 2% wettable powder	1977	Laboratory	1% powder: death; time to death ranged from no effect to 72 hr 2% powder: death; time to death range 24–48 hr	Kihara & Yamashita (1978)
Cyhexatin	*Eumeces marginatus*, Ousima skink	1% and 2% wettable powder	1977	Laboratory	1% powder: no effect 2% powder: death; time to death ranged from no effect to 72 hr	Kihara & Yamashita (1978)

Table 4. (Continued).

Contaminant	Species	Exposure	Year	Location	Effects	Reference
DAEP	*Eumeces marginatus*, Ousima skink	1% and 2% emulsifiable concentration	1977	Laboratory	1% concentration: death; time to death range 12–48 hr. 2% concentration: death; time to death range 1–3 hr	Kihara & Yamashita (1978)
DDD or TDE	*Cnemidophorus tessellatus*, checkered whiptail	Daily early morning application in cotton fields during summer months	1967	Presidio Basin, Rio Grande Valley, Southwest Texas, USA	Distribution, behavior, reproduction, growth rate, and population structure is not affected	Saxon (1970)
DDE	*Cnemidophorus tessellatus*, checkered whiptail	Daily early morning application in cotton fields during summer months	1967	Presidio Basin, Rio Grande Valley, Southwest Texas, USA	Distribution, behavior, reproduction, growth rate, and population structure is not affected	Saxon (1970)
DDT	*Eumeces marginatus*, Ousima skink	1% and 2% concentration, technical product	1977	Laboratory	Both concentrations: death; time to death 48 hr	Kihara & Yamashita (1978)
DDT	*Cnemidophorus tessellatus*, checkered whiptail	Daily early morning application in cotton fields during summer months	1967	Presidio Basin, Rio Grande Valley, Southwest Texas, USA	Distribution, behavior, reproduction, growth rate, and population structure are not affected	Saxon (1970)

Table 4. (Continued).

Contaminant	Species	Exposure	Year	Location	Effects	Reference
Diazinon	*Eumeces marginatus*, Ousima skink	1% and 2% emulsifiable concentration and 3% dust	1977	Laboratory	1% concentration: death; time to death 1 hr 2% concentration: death; time to death range 1–3 hr 3% dust: death; time to death range 3–72 hr	Kihara & Yamashita (1978)
Dicofol	*Eumeces marginatus*, Ousima skink	1% and 2% emulsifiable concentration and 3% dust	1977	Laboratory	1% concentration: death; time to death ranged from no effect to 12 hr 2% concentration: death; time to death range 6–12 hr 3% dust: no effect	Kihara & Yamashita (1978)
Dieldrin	*Eumeces marginatus*, Ousima skink	1% and 2% concentration, technical product	1977	Laboratory	1% concentration: no effect 2% concentration: death; time to death 24 hr	Kihara & Yamashita (1978)
Estradiol benzoate	*Eublepharis macularius*, leopard gecko	0.1 μg estradiol/5 μL ethanol and 1.0 μg estradiol/5 μL ethanol injected into eggs at day 5 and day 11 of development	Early 1990s	Laboratory	Both doses: decreased % of male hatchlings observed for eggs injected at both day 5 and day 11 (no differences between days)	Tousigant & Crews (1994)

Table 4. (Continued).

Contaminant	Species	Exposure	Year	Location	Effects	Reference
Estradiol benzoate	*Eublepharis macularius*, leopard gecko	10 μg estradiol/5 μL ethanol injected into eggs at day 5 and day 11 of development	Early 1990s	Laboratory	Eggs injected at both day 5 and day 11: significant decrease in % of male hatchlings observed (no differences between days) Day 11 injected eggs: significantly higher mortality observed Overall results: decrease in length of incubation with increasing dose was observed for eggs injected at day 5; increase in length of incubation with increasing dose was observed for eggs injected at day 11; no growth differences	Tousigant & Crews (1994)
Estradiol benzoate	*Eublepharis macularius*, leopard gecko	10, 15, or 20 μg of powdered 17β-estradiol benzoate dissolved/μL of ethanol injected into eggs	1986–1987	Laboratory	Sex reversal: developed ovaries instead of testes	Bull et al. (1988)

Table 4. (Continued).

Contaminant	Species	Exposure	Year	Location	Effects	Reference
					1986 results with 5 µL of 15 or 20 µg/µL estradiol: 3–6 d after egg laying, 67% female; 1–14 d after egg laying, 19% female 1987 results with 5 µg/µL estradiol: 3–6 d after egg laying, 100% female	
Ethylnitrophenyl phenylphosphonothiate (EPN)	*Eumeces marginatus*, Ousima skink	1% and 2% emulsifiable concentration and 1.5% dust	1977	Laboratory	1% concentration: death; time to death range 3–6 hr 2% concentration: death; time to death 3 hr 1.5% dust: death; time to death range 6–48 hr	Kihara & Yamashita (1978)
Fadrozole	*Cnemidophorus uniparens*, desert grassland whiptail	Single treatment of 100 µg of fadrozole dissolved in 5 µL of 95% ethanol applied topically to vascularized portion of eggs laid by parthenogenic females between day of lay to 11 d after lay	Early 1990s	Laboratory	Male sex determination induced in all eggs	Wibbels & Crews (1994)

Table 4. (Continued).

Contaminant	Species	Exposure	Year	Location	Effects	Reference
Fenitrothion (or MEP)	*Eumeces marginatus*, Ousima skink	1% and 2% emulsifiable concentration	1977	Laboratory	1% concentration: death; time to death range 24–48 hr 2% concentration: death; time to death 3 hr	Kihara & Yamashita (1978)
Fenobucarb	*Eumeces marginatus*, Ousima skink	2% dust	1977	Laboratory	Death; time to death range 24–72 hr	Kihara & Yamashita (1978)
Fentin hydroxide	*Eumeces marginatus*, Ousima skink	1% and 2% concentration, technical product	1977	Laboratory	1% concentration: death; time to death ranged from no effect to 48 hr 2% concentration: death; time to death range 3–12 hr	Kihara & Yamashita (1978)
Lindane	*Eumeces marginatus*, Ousima skink	1% and 2% concentration, technical product	1977	Laboratory	1% concentration: death; time to death ranged from no effect to 72 hr 2% concentration: death; time to death 24 hr	Kihara & Yamashita (1978)
MAFe	*Eumeces marginatus*, Ousima skink	1% and 2% concentration, technical product	1977	Laboratory	1% concentration: death; time to death range 6–12 hr 2% concentration: death; time to death range 6–24 hr	Kihara & Yamashita (1978)

Table 4. (Continued).

Contaminant	Species	Exposure	Year	Location	Effects	Reference
MBCP (or leptophos)	*Eumeces marginatus*, Ousima skink	1% and 2% emulsifiable concentration	1977	Laboratory	1% concentration: death; time to death 1 hr 2% concentration: death; time to death range 6–12 hr	Kihara & Yamashita (1978)
MPMC (or xylylcarb)	*Eumeces marginatus*, Ousima skink	2% dust	1977	Laboratory	Death; time to death ranged from no effect to 12–72 hr	Kihara & Yamashita (1978)
MPP (or fenthion)	*Eumeces marginatus*, Ousima skink	1% and 2% emulsifiable concentration	1977	Laboratory	1% concentration: no effect 2% concentration: death; time to death ranged from no effect to 72 hr	Kihara & Yamashita (1978)
MTMC (or metolcarb)	*Eumeces marginatus*, Ousima skink	2% dust	1977	Laboratory	Death; time to death ranged from no effect to 6–72 hr	Kihara & Yamashita (1978)
Malathion	*Lacerta parva*, dwarf lizard	1 mg/kg, 2 mg/kg, and 3 mg/kg doses (96% technical purity) administered daily for 16 wk in the laboratory	1991	Collected near Tokathan and Harmandal, near Eskişhir, Turkey	All three doses: heavy lipid accumulation around liver, kidney, intestines, and liver; slight kidney and intestinal damage occurred Mean LD_{50} = 169.8 mg/kg	Özelmas & Akay (1995)

Table 4. (Continued).

Contaminant	Species	Exposure	Year	Location	Effects	Reference
Malathion	*Anolis carolinensis*, green anole	99% dissolved in corn oil, dose administered in laboratory and equivalent to 5% body weight of each lizard	1980	Captured from Merritt Island National Wildlife Refuge, Florida and in Gainesville, Florida, USA	LD_{50} = 2,324 mg/kg, 95% confidence interval: 1,671–3,234 mg/kg Levels of cholinesterase activity (inhibition) as % of normal related to 50% mortality: 44.4%	Hall & Clark (1982)
Malathion	*Eumeces marginatus*, Ousima skink	1% and 2% emulsifiable concentration and 3% dust	1977	Laboratory	1% concentration: no effect 2% concentration: death; time to death range 3–12 hr 3% dust: death; time to death ranged from no effect to 48–72 hr	Kihara & Yamashita (1978)
Meldrin	*Eumeces marginatus*, Ousima skink	1% and 2% emulsifiable concentration	1977	Laboratory	1% concentration: death; time to death range 3–6 hr 2% concentration: death; time to death range 6–24 hr	Kihara & Yamashita (1978)
Methaldehyde	*Eumeces marginatus*, Ousima skink	3.2% and 6% dust	1977	Laboratory	No effect at either dose	Kihara & Yamashita (1978)

Table 4. (Continued).

Contaminant	Species	Exposure	Year	Location	Effects	Reference
Methomyl	*Eumeces marginatus*, Ousima skink	1% and 2% wettable powder	1977	Laboratory	1% powder: death; time to death 3 hr 2% powder: death; time to death range 1–3 hr	Kihara & Yamashita (1978)
Methyl parathion	*Anolis carolinensis*, green anole	98% dissolved in corn oil, dose administered in laboratory and equivalent to 5% body weight of each lizard	1980	Captured from Merritt Island National Wildlife Refuge, Florida and in Gainesville, Florida, USA	LD_{50} = 82.7 mg/kg, 95% confidence interval: 56.2–187.9 mg/kg Levels of cholinesterase activity (inhibition) as % of normal related to 50% mortality: 51.4%	Hall & Clark (1982)
Methyl parathion	*Cnemidophorus tesselatus*, checkered whiptail	Daily early morning application in cottonfields during summer months	1967	Presidio Basin, Rio Grande Valley, Southwest Texas, USA	Distribution, behavior, reproduction, growth rate, and population structure are not affected	Saxon (1970)
Nicotine sulfate	*Eumeces marginatus*, Ousima skink	1% and 2% emulsifiable concentration	1977	Laboratory	1% concentration: death; time to death ranged from no effect to 72 hr 2% concentration: death; time to death range 12–24 hr	Kihara & Yamashita (1978)

Table 4. (Continued).

Contaminant	Species	Exposure	Year	Location	Effects	Reference
Oxine Copper	*Eumeces marginatus*, Ousima skink	1% and 2% concentration, technical product	1977	Laboratory	1% concentration: death; time to death ranged from no effect to 24 hr. 2% concentration: death; time to death 12 hr	Kihara & Yamashita (1978)
Parathion	*Gallotia galloti*, Gallot's lizard	1 dose of 0.5, 2.5, or 5.0 mg active ingredient/kg body mass	1996	Collected on Tenerife, Canary Islands, Spain	All doses: no mortality. Brain, serum, and liver microsomal esterase activity and liver microsomal monooxygenase activities inhibited 24 hr after single acute treatment. Serum butyrylcholinesterase carboxylesterase: suitable nondestructive biomarker of effect	Sanchez et al. (1997)
Parathion	*Gallotia galloti*, Gallot's lizard	1 dose of 7.5 mg active ingredient/kg body mass	1996	Collected on Tenerife, Canary Islands, Spain	No mortality. Brain, serum, and liver microsomal esterase activity and liver microsomal monooxygenase inhibited 24 hr after single acute treatment	Sanchez et al. (1997)

Table 4. (Continued).

Contaminant	Species	Exposure	Year	Location	Effects	Reference
					Except for brain esterase activity after 6 hr, inhibition of brain acetylcholinesterase indicates poisoning Serum butyrylcholinesterase carboxylesterase: suitable nondestructive biomarker of effect	
Parathion	*Gallotia galloti*, Gallot's lizard	Three consecutive doses over 72 d of 0.5 or 7.5 mg active ingredient/kg body mass	1996	Collected on Tenerife, Canary Islands, Spain	No mortality occurred as a result of either dose Inhibition of brain acetylcholinesterase indicates poisoning; increase in recovery time to normal esterase activity observed after each consecutive acute treatment Serum butyrylcholinesterase carboxylesterase: suitable nondestructive biomarker of effect	Sanchez et al. (1997)

Table 4. (Continued).

Contaminant	Species	Exposure	Year	Location	Effects	Reference
Parathion	*Anolis carolinensis*, green anole	98% dissolved in corn oil, dose administered in laboratory and equivalent to 5% body weight of each lizard	1980	Captured from Merritt Island National Wildlife Refuge, Florida and in Gainesville, Florida, USA	LD_{50} = 8.9 mg/kg, 95% confidence interval: 4.7–13.2 mg/kg Levels of cholinesterase activity (inhibition) as % of normal related to 50% mortality: 22.7%	Hall & Clark (1982)
Parathion	*Cnemidophorus tessellatus*, checkered whiptail	Daily early morning application in cottonfields during summer months	1967	Presidio Basin, Rio Grande Valley, Southwest Texas, USA	Distribution, behavior, reproduction, growth rate, and population structure are not affected	Saxon (1970)
Pentachlorophenol	*Eumeces marginatus*, Ousima skink	2% concentration, technical product	1977	Laboratory	1% concentration: death; time to death range 3–6 hr 2% concentration: death; time to death 3 hr	Kihara & Yamashita (1978)

Table 4. (Continued).

Contaminant	Species	Exposure	Year	Location	Effects	Reference
Phenthoate	*Eumeces marginatus*, Ousima skink	2% dust	1977	Laboratory	Death; time to death ranged from no effect to 72 hr	Kihara & Yamashita (1978)
Propoxur	*Eumeces marginatus*, Ousima skink	1% dust	1977	Laboratory	Death; time to death ranged from no effect to 48 hr	Kihara & Yamashita (1978)
Prothiophos	*Eumeces marginatus*, Ousima skink	1% and 2% emulsifiable concentration	1977	Laboratory	1% concentration: death; time to death range 1–24 hr 2% concentration: death; time to death range 12–72 hr	Kihara & Yamashita (1978)
Pyrethrin	*Eumeces marginatus*, Ousima skink	1% and 2% emulsifiable concentration	1977	Laboratory	1% concentration: death; time to death 1 hr 2% concentration: death; time to death 3 hr	Kihara & Yamashita (1978)
Quinoxaline	*Eumeces marginatus*, Ousima skink	1% and 2% wettable powder	1977	Laboratory	1% powder: death; time to death ranged from no effect to 72 hr 2% powder: death; time to death 72 hr	Kihara & Yamashita (1978)
Rotenone	*Eumeces marginatus*, Ousima skink	3% dust	1977	Laboratory	Death; time to death range 12–72 hr	Kihara & Yamashita (1978)

Table 4. (Continued).

Contaminant	Species	Exposure	Year	Location	Effects	Reference
Sodium fluoroacetate (compound 1080)	*Tiliqua rugosa*, shingleback skink	Administered intraperitoneally in aqueous solution in laboratory at doses of 50, 100, 200, 400, and 800 mg/kg	Early 1980s	Populations from Kulikup and Rawlinna, West Australia and South Australia	Death: Kulikup, West Australian populations: LD_{50} = >800 mg/kg; Rawlinna, West Australian populations: LD_{50} = 200 mg/kg; South Australian populations: LD_{50} = 200 mg/kg	Twigg & Mead (1990)
Sodium fluoroacetate	*Tiliqua rugosa*, shingleback skink	One treatment of 25, 100, or 250 mg/kg in the laboratory	1983–1984	Populations from Kulikup, West Australia with evolutionary exposure to fluoroacetate	Effects on male fecundity: 10%–80% of tubules degenerated in testes after 1 treatment of 250 mg/kg	Twigg et al. (1988)
Sodium fluoroacetate	*Tiliqua rugosa*, shingleback skink	Five treatments of 5, 20, and 50 mg/kg in the laboratory	1983–1984	Populations from Kulikup, West Australia with evolutionary exposure to fluoroacetate	Effects on male fecundity: 10%–50% of tubules degenerated in testes after five treatments of 50 mg/kg	Twigg et al. (1988)
Sodium fluoroacetate	*Pogona barbata*, bearded dragon	92% and 95.8% purity at doses of 10 and 100 mg/mL in laboratory	Early 1980s	Collected from Brindabella Range, New South Wales, Australia	Time until effects (lethargic, no movement, foaming at mouth) = 15.2 hr Time until death = 22.8 hr (range 14.9–24.2 hr)	McIlroy et al. (1985)

Table 4. (Continued).

Contaminant	Species	Exposure	Year	Location	Effects	Reference
Sodium fluoroacetate	*Varanus gouldii*, sand monitor	92% and 95.8% purity at doses of 25 and 50 mg/mL in laboratory	Early 1980s	Collected from Yunta, South Australia	LD_{50} = <110 mg/kg (a dose of 110–450 mg/kg caused death) One animal survived a dose of 50 mg/kg Time until effects (lethargic, no movement, foaming at mouth) = 47.4 hr (range 24.2–141.2 hr) Time until death = 111.6 hr (range 66.5–292.5 hr) LD_{50} = 43.6 mg/kg (95% CL = 27.5–69.2 hr) Two animals did recover	McIlroy et al. (1985)
Sodium fluoroacetate	*Varanus varius*, lace monitor	92% and 95.8% purity at dose of 100 mg/mL in laboratory	Early 1980s	Collected from Macquaries Marshes, Australia	Time until effects (lethargic, no movement, foaming at mouth) = 83.9 hr (range 26.6–141.3 hr) Time until death = 109.5 hr (range 73.6–145.4 hr) LD_{50} = <119 mg/kg (a dose of 119 and 150 mg/kg caused death)	McIlroy et al. (1985)

Table 4. (Continued).

Contaminant	Species	Exposure	Year	Location	Effects	Reference
Sodium fluoroacetate	*Tiliqua nigrolutea*, blotched blue-tongue skink	92% and 95.8% purity at dose of 100 mg/mL in laboratory	Early 1980s	Collected from Brindabella Range, New South Wales, Australia	Time until effects (lethargic, no movement, foaming at mouth) = 90.6 hr (range 13.3–160.9 hr) Time until death = 130.5 hr (range 14.4–522.5 hr) LD_{50} = 336.4 mg/kg (95% CL = 232.4–487.1 hr)	McIlroy et al. (1985)
Sodium fluoroacetate	*Tiliqua rugosa*, shingleback skink	92% and 95.8% purity at doses of 100 and 400 mg/mL in laboratory	Early 1980s	Collected from South Australia	Time until death = 89.0 hr (range 21.0–134.0 hr) LD_{50} = 205.9 mg/kg (95% CL = 147.2–289.1 hr)	McIlroy et al. (1985)
Sodium fluoroacetate	*Tiliqua rugosa*, shingleback skink	92% and 95.8% purity at doses of 100 and 400 mg/mL in laboratory	Early 1980s	Collected from the southwest of Western Australia	Time until death = 174.0 hr (range 22.0–363.0 hr) LD_{50} = 507.7 mg/kg (95% CL = 447.0–577.1 hr)	McIlroy et al. (1985)

Table 4. (Continued).

Contaminant	Species	Exposure	Year	Location	Effects	Reference
Sodium fluoroacetate	*Tiliqua rugosa*, shingleback skink	92% and 95.8% purity at doses of 100, 200, and 800 mg/mL in laboratory	Early 1980s	Collected from the southwest of Western Australia	Time until death = 168.0 hr (range 24.0–432.0 hr) LD_{50} = 543.2 mg/kg (95% CL = 500.5–589.5 hr)	McIlroy et al. (1985)
Trichlorfon	*Eumeces marginatus*, Ousima skink	1% and 2% emulsifiable concentration and 3% dust	1977	Laboratory	1% concentration: no effect; 2% concentration: death; time to death range 6–24 hr; 3% dust: death; time to death ranged from no effect to 48 hr	Kihara & Yamashita (1978)

lethal effects data and 9 contained sublethal effects information. Two studies on the effects of inorganic contaminants on lizards and 20 studies on the lethal and sublethal effects of radionuclides/radiation on lizards (Table 5) were reviewed.

A. Organic Contaminants

The lethal effects of pesticides were available for 9 species, representing five families (Table 1). The sublethal effects of organic contaminants were available for seven families of lizards, which included 11 species (Table 1).

Hall and Clark (1982) used green anoles collected in Florida to estimate the lethal dose for 50% mortality (LD_{50}) for azinphos-methyl, malathion, methyl parathion, and parathion (Table 4). Mortality of 80%–100% occurred in green anoles at doses of 12.2 mg/kg parathion, 83 mg/kg methyl parathion, 139 mg/kg azinphos-methyl, and 3,000 mg/kg malathion administered as stomach poisons. They found that green anole brain cholinesterase activity was related to doses of azinphos-methyl, malathion, methyl parathion, and parathion; 50% inhibition of brain cholinesterase was associated with death, whereas 40% inhibition indicated sublethal exposure. Hall and Clark (1982) concluded that responses of green anoles were similar to those observed for mallard ducks and rats.

Saxon (1970) concluded that the distribution, behavior, reproduction, growth rate, and population structure of checkered whiptails in Southwest Texas (Presidio Basin, Rio Grande Valley) was not affected by DDT, DDE, DDD, methyl parathion, or parathion. Direct exposure to the pesticides was diminished because the daily activity pattern of the checkered whiptails did not coincide with early morning pesticide applications to cotton fields during the summer.

Kihara and Yamashita (1978) studied the lethal effects of 34 different pesticides on Ousima skinks. Time of death was estimated for different exposures of the pesticides (Table 4).

In a laboratory study using Gallot's lizards from Tenerife, Canary Islands, Spain, brain, serum, and liver microsomal esterase activity and liver microsomal monooxygenase activities were inhibited by various doses of parathion (Sanchez et al. 1997) (Table 4). A nonlinear correlation was found between the destructive biomarker brain acetylcholinesterase and the nondestructive biomarker serum butyrylcholinesterase. No lizards died as a result of parathion exposure. Sanchez et al. (1997) concluded that Gallot's lizards were ideal bioindicators. Similar results were obtained in an earlier study by Fossi et al. (1995) in which Gallot's lizards collected from Tenerife were exposed to various doses of trichlorfon (reviewed in Lambert 1997b).

Dwarf lizards collected near Eskişehir, Turkey, exhibited liver, kidney, and intestinal damage as a result of three low doses of malathion administered in the laboratory (Özelmas and Akay 1995). In addition, heavy lipid accumulation around the liver, kidney, and intestines was observed. For malathion, they calculated an LD_{50} of 169.8 mg/kg for dwarf lizards.

Three laboratory studies were conducted to determine the sensitivity of Aus-

Table 5. Lizard Lethal and Sublethal Effects Data for Radionuclides/Radiation.

Contaminant	Species	Exposure	Year	Location	Effects	Reference
^{137}Cesium	*Uta stansburiana*, side-blotched lizard	Continuous ^{137}Cs source (33,000 Ci) since Jan. 1964, ~750 rads/yr (range 360–1,800 rads) or 2 rads/d	1975	9-ha site, Nevada Test Site, Nye County, Nevada, USA	Radiation-sterilized females were heavier (as compared to nonirradiated controls) because they had extraordinarily large fat storage deposits. Sterile females had much lower rates of energy expenditure via respiration and lower rates of energy intake by feeding (indirect responses to radiation-induced sterility)	Nagy & Medica (1985)
^{137}Cesium	*Uta stansburiana*, side-blotched lizard	Continuous ^{137}Cs source (33,000 Ci) since Jan. 1964, ~1.5–10 R/d	1975–1977	9-ha site, Nevada Test Site, Nye County, Nevada, USA	Direct radiation damage to gametogenesis and the gonads cause sterility. Older males were also sterile. Pituitary function does not appear to be damaged by radiation	Pearson et al. (1978)

Table 5. (Continued).

Contaminant	Species	Exposure	Year	Location	Effects	Reference
^{137}Cesium	*Uta stansburiana*, side-blotched lizard	Continuous ^{137}Cs source (33,000 Ci) since Jan. 1964, ~750 rads/yr (range 360–1,800 rads) or 2 rads/d	1964–1973	9-ha site, Nevada Test Site, Nye County, Nevada, USA	Females became sterile as early as 11 mon of age (but after the lizard reproduced in its first breeding season). Sterility may occur at any time after 12 mon; some were sterile at 20 mon and others were still reproductive. Almost all females 32 mon of age or older were sterile. Some females were sterile after accumulated doses of about 500 rads, while others may have required 1,000 or more rad. By 1968, there was evidence of reproductive failure. Fat bodies were enormously hypertrophied. The population can withstand chronic radiation stress because yearling lizards (~75% of spring population) reproduce normally before damaging doses are accumulated	Medica et al. (1973); Turner (1975); Turner & Medica (1977)

Table 5. (Continued).

Contaminant	Species	Exposure	Year	Location	Effects	Reference
^{137}Cesium	*Callisaurus draconoides*, zebra-tailed lizard	Continuous ^{137}Cs source (33,000 Ci) since Jan. 1964, ~400–500 rads/yr	1964–1971	9-ha site, Nevada Test Site, Nye County, Nevada, USA	Females have become sterile (by 1969); the absence of ovarian tissue and the hypertrophy of fat bodies was also observed	Turner (1975)
^{137}Cesium	*Gambelia wislizenii*, longnose leopard lizard	Continuous ^{137}Cs source (33,000 Ci) since Jan. 1964, ~400–500 rads/yr	1964–1971	9-ha site, Nevada Test Site, Nye County, Nevada, USA	All females have become sterile (by 1969); the absence of ovarian tissue and the hypertrophy of fat bodies was also observed. An accumulated dose of about 1,500 rads was sufficient to destroy the ovary. The species eventually became extinct from the irradiated plot	Turner et al. (1973); Medica et al. (1973); Turner (1975); Turner & Medica (1977)
^{137}Cesium	*Cnemidophorus tigris*, western whiptail	Continuous ^{137}Cs source (33,600 Ci) since Jan. 1964, ~200–250 rads/yr	1964–1971	9-ha site, Nevada Test Site, Nye County, Nevada, USA	Some females have become sterile (starting in 1969); the absence of ovarian tissue and the hypertrophy of fat bodies was also observed. An accumulated dose of about 1,500 rads was sufficient to destroy the ovary. There is no evidence that the population is going extinct	Turner et al. (1973); Medica et al. (1973); Turner (1975); Turner & Medica (1977)

Table 5. (Continued).

Contaminant	Species	Exposure	Year	Location	Effects	Reference
^{137}Cesium	*Phrynosoma platyrhinos*, desert horned lizard	Continuous ^{137}Cs source (33,600 Ci) since Jan. 1964, ~400–500 rads/yr	1964–1971	9-ha site, Nevada Test Site, Nye County, Nevada, USA	A decline in population density was observed, and the population is becoming extinct. Reproductive failure has occurred (probably started ~1966–1967). All females are sterile due to the absence of ovarian tissue and the hypertrophy of fat bodies	French (1970); Medica et al. (1973); Turner (1975)
^{137}Cesium	*Uta stansburiana*, side-blotched lizard	Continuous ^{137}Cs source (33,000 Ci) since Jan. 1964, ~750 rads/yr (range 360–1,800 rads) or 2 rads/d	1965–1968	9-ha site, Nevada Test Site, Nye County, Nevada, USA	No evident changes in the dispersion of the population (reorientation) occupying the irradiated exposure were observed	Turner (1975)
^{137}Cesium	*Uta stansburiana*, side-blotched lizard	Continuous ^{137}Cs source (33,000 Ci) since Jan. 1964, ~750 rads/yr (range 360–1,800 rads) or 2 rads/d	1964–1968	9-ha site, Nevada Test Site, Nye County, Nevada, USA	No statistically significant difference in sex ratios, maximum life span, or age distributions (as compared with control populations) was observed. Four years of 2 rads/d apparently has no effect	Turner et al. (1969); French (1970)
^{137}Cesium	*Uta stansburiana*, side-blotched lizard	Continuous ^{137}Cs source (33,600 Ci) since Jan. 1964, ~1,800–2,000 rads/yr or 250 rads/mon	1965–1966	9-ha site, Nevada Test Site, Nye County, Nevada, USA	No acute lethal effect was observed. Delayed effects are unknown	Turner & Lannom (1968)

Table 5. (Continued).

Contaminant	Species	Exposure	Year	Location	Effects	Reference
^{137}Cesium	*Gambelia wislizenii*, longnose leopard lizard	Continuous ^{137}Cs source (33,600 Ci) July–Nov. 1964, ~10 R/d for 125 d or 1,000–2,000 R for 5 mon	1965–1966	9-ha site, Nevada Test Site, Nye County, Nevada, USA	No acute lethal effect was observed. Delayed effects are unknown	Turner & Lannom (1968)
^{137}Cesium	*Cnemidophorus tigris*, western whiptail	Continuous ^{137}Cs source (33,600 Ci) since Jan. 1964, ~400–600 rads/yr	1965–1966	9-ha site, Nevada Test Site, Nye County, Nevada, USA	No acute lethal effect was observed. Delayed effects are unknown	Turner & Lannom (1968)
^{137}Cesium	*Uta stansburiana*, side-blotched lizard	Continuous ^{137}Cs source (33,600 Ci) since Jan. 1964, ~1,800–2,000 rads/yr or 250 rads/mon	1964	9-ha site, Nevada Test Site, Nye County, Nevada, USA	A decrease in growth in the irradiated plot may be caused by radiation (could be density dependent)	Turner et al. (1965)
^{137}Cesium	*Anolis gundlachi*, Gundlach's anole	^{137}Cs source (10,000 Ci) for 3 mon; cumulative free-air dose at 15 m was ~13 kR (rate of ~140 R/d)	1963–1965	0.25-ha site, El Verde, Puerto Rico	A zone of almost complete lethality extended 15–20 m from the source (~200–250 adults were killed). Lizards received a chronic dose of 84 rads/d (which was lethal), resulting in a cumulative dose of ~7,800 rads. Young lizards experienced better survival (because of time spent below ground and entering the population during the experiment)	Turner & Gist (1970)

Table 5. (Continued).

Contaminant	Species	Exposure	Year	Location	Effects	Reference
137Cesium (continued)	*Anolis gundlachi*, Gundlach's anole	137Cs source (10,000 Ci) for 3 mon; cumulative free-air dose at 15 m was ~13 kR (rate of ~140 R/d)	1963–1965	0.25-ha site, El Verde, Puerto Rico	Irradiation possibly caused altered perch heights among males. In addition, large males experienced better survival as evidenced by changes in postirradiation size distributions. Two species of anoles (*Anolis cristatellus*, crested anole and *Anolis krugi*, Krug's anole), which are found only in open areas, invaded the irradiated area because of the loss of forest canopy leaves	Turner & Gist (1970)
137Cesium	*Anolis evermanni*, Evermann's anole	137Cs source (10,000 Ci) for 3 mon; cumulative free-air dose at 15 m was ~13 kR (rate of ~140 R/d)	1963–1965	0.25-ha site, El Verde, Puerto Rico	A zone of almost complete lethality extended 15–20 m from the source. Lizards received a chronic dose of 84 rads/d (which was lethal), resulting in a cumulative dose of ~7,800 rads. Two species of anoles (*Anolis cristatellus*, crested anole and *Anolis krugi*, Krug's anole), which are found only in open areas, invaded the irradiated area because of the loss of forest canopy leaves	

Table 5. (Continued).

Contaminant	Species	Exposure	Year	Location	Effects	Reference
^{60}Cobalt	*Uta stansburiana*, side-blotched lizard	Young and adults from the Nevada Test Site were exposed to a dose from 635 R to 3,600 R administered at 100 R/min and 200 R/min. Irradiation was from a 10,000-Ci ^{60}Co source.	1963–1965	Laboratory	Dose of 635–1,200 R; no difference in survival or growth between irradiated and control animals was observed. Dose of 1,200–1,600 R; survival was unimpaired but growth was retarded. Dose >1,600 R–2,400 R; survival and growth were lower than those of controls. $LD_{50/30}$ (50% mortality within 30 d) = 1,700–2,200 R. Adult males were more radiosensitive than adult females, but no sex differences were observed in an experiment with lizards 3–4 mon old	Turner et al. (1965); Turner et al. (1967)
^{60}Cobalt	*Scincella lateralis*, ground skink	Received doses of 300, 600, 900, 1,200, and 1,500 R at a dose rate of 99.7 R in the laboratory from a ^{60}Co source.	1961	Collected from University of Florida campus, Alachua County, Gainesville, Florida, USA	One skink that received a dose of 600 R died after 16 d; one that received a dose of 900 R died after 16 d; and one that received a dose of 1,500 R died after 13 d. The results indicate that ground skinks are very resistant to gamma radiation. Because the dosages did not kill many skinks, a $LD_{50/30}$ could not be calculated	Brooks (1962)

Table 5. (Continued).

Contaminant	Species	Exposure	Year	Location	Effects	Reference
X-Irradiation	*Calotes versicolor*, variable agama	Eggs were exposed to one dose of 10,250 rads (A), one dose of 20,500 rads (B), and two doses of 10,250 (total dose of 20,500 rads) (C)	Early 1970s	Laboratory	Experiment A: Embryos observed 1 d after irradiation. Survival was 87.5%. Experiment B: Embryos observed 2 d after irradiation. Survival was 67.2%. Experiment C: Doses were 2 d apart. Embryos observed 4 d after first dose. Survival was 66.7%. In all experiments, irradiation caused mortality, retardation in development and growth rate, hemorrhage, axial defects, unilateral microphthalmia, and eye damage	Chiplonkar & Goel (1975)
X-Irradiation	*Hemidactylus leschenaulti*, Leschenault's leaf-toed gecko	Whole-body doses of 1,000 R	~1971	Laboratory	X-irradiation alters the amount of esterases in the kidneys (increased) and the lungs (decreased) but not in the brain or liver. Metabolic alterations could eventually lead to pathological changes or even death	George & Eapen (1973)

Table 5. (Continued).

Contaminant	Species	Exposure	Year	Location	Effects	Reference
X-Irradiation	*Uta stansburiana*, side-blotched lizard	Dose: 70 R/min, 450 R of gonadal radiation before breeding season	1962	Two areas of 0.84 ha (similar habitat), 3.2 km apart (Area I, treatment area, established May 1960; Area II, control site, established May 1961), Western Texas, USA.	A 50% decline in natality (number of young produced) resulted in a striking decrease in density the year after irradiation. The decline may be a result of temporary sterility or a higher rate of mutations affecting embryonic survival. The breeding structure of the populations was affected through interference by irradiation with normal selective processes in a natural population. The results were the same in a switchback experiment performed in 1964	Tinkle (1965)
X-Irradiation	*Uta stansburiana*, side-blotched lizard	Whole-body doses ranging from 250 R to 3,000 R	1961–1962	Laboratory	$LD_{50/30}$ = 1,000–1,200 R. Temporary sterilization was produced by a whole-body dose of 500 R, and permanent sterilization occurred with whole-body doses >800 R	Dana & Tinkle (1965)
X-Irradiation	*Anolis carolinensis*, green anole	Whole-body doses of 450–500 R/min until death.	1958	Laboratory	Death after receiving a dose of 107,000 R at 5 °C or 108,000 R at 37°C	Smith & Thomson (1959)

Table 5. (Continued).

Contaminant	Species	Exposure	Year	Location	Effects	Reference
Thermonuclear detonation	*Uta stansburiana*, side-blotched lizard	Project Sedan, 100 ± 15 kiloton device buried 193.6 m	1962–1963	Yucca Flat, Nevada Test Site, Nye County Nevada, USA	Lizards were exterminated to a distance of 1,219 m from ground zero (attributable to physical effects of detonation). No changes due to detonation were detected at 2,591–2,743 m. Delayed mortality of juveniles hatched after the test attributed to destruction of habitat. Residual gamma radiation was not likely to be lethal	Turner & Gist (1965)
Thermonuclear detonation	*Gambelia wislizenii*, longnose leopard lizard	Project Sedan, 100 ± 15 kiloton device buried 193.6 m	1962–1963	Yucca Flat, Nevada Test Site, Nye County Nevada, USA	Lizards were exterminated to a distance of 1,219 m from ground zero (attributable to physical effects of detonation). No changes due to detonation were detected at 2,591–2,743 m. Residual gamma radiation was not likely to be lethal	Turner & Gist (1965)

Table 5. (Continued).

Contaminant	Species	Exposure	Year	Location	Effects	Reference
Thermonuclear detonation	*Cnemidophorus tigris*, western whiptail	Project Sedan, 100 ± 15 kiloton device buried 193.6 m	1962–1963	Yucca Flat, Nevada Test Site, Nye County Nevada, USA	Lizards were exterminated to a distance of 1,219 m from ground zero (attributable to physical effects of detonation). No changes due to detonation were detected at 2,591–2,743 m. This species appeared to be more sensitive as fewer adults were observed between 1,372 and 1,524 m from ground zero. Residual gamma radiation was not likely to be lethal	Turner & Gist (1965)
Natural radiation	*Uta stansburiana*, side-blotched lizard	Natural areas with high surface radiation of 25–100 mR/hr, mined for radium, vanadium, and uranium	1959–1964	Natural undisturbed areas (not mined areas) in the Upper Colorado River Basin, Utah, USA (Emery County, Grand County, and San Juan County)	There was no significant variation in external morphology between lizards that lived in areas of high surface radiation as compared to areas of low surface radiation	Tanner (1965)

Table 5. (Continued).

Contaminant	Species	Exposure	Year	Location	Effects	Reference
Natural radiation	*Sceloporus undulatus elongatus*, northern plateau lizard	Natural areas with high surface radiation of 25–100 mR/hr, mined for radium, vanadium, and uranium	1959–1964	Natural undisturbed areas (not mined areas) in the Upper Colorado River Basin, Utah, USA (Emery County, Grand County, and San Juan County)	There was no significant variation in external morphology between lizards that lived in areas of high surface radiation as compared to areas of low surface radiation	Tanner (1965)
Natural radiation	*Cnemidophorus tigris septentrionalis*, Painted Desert whiptail	Natural areas with high surface radiation of 25–100 mR/hr, mined for radium, vanadium, and uranium	1959–1964	Natural undisturbed areas (not mined areas) in the Upper Colorado River Basin, Utah, USA (Emery County, Grand County, and San Juan County)	There was no significant variation in external morphology between lizards that lived in areas of high surface radiation as compared to areas of low surface radiation	Tanner (1965)

tralian lizards to sodium fluoroacetate (compound 1080), a mammalian pesticide (McIlroy et al. 1985; Twigg et al. 1988; Twigg and Mead 1990). Time of death and LD_{50}s were estimated for five lizard species that have been observed or would be likely to scavenge poisoned meat or vegetable baits (McIlroy et al. 1985; Twigg and Mead 1990). Populations of shingleback skinks that coexist with fluoroacetate-bearing vegetation in parts of western Australia were much less sensitive to poisoning than those not exposed to the toxic plants (Table 4). The time until effects were observed (lethargic, no movement, foaming at mouth) was estimated by McIlroy et al. (1985) for four species observed eating or likely to eat poison baits (Table 4). Effects on male fecundity in shingleback skinks with evolutionary exposure to fluoroacetate were observed in a study by Twigg et al. (1988) (Table 4). A single dose of 250 mg/kg or five treatments of 50 mg/kg administered during 12 d resulted in the degeneration of seminiferous tubules.

Exogenous estradiol (or estrogen) caused female development in leopard geckos, a species in which sex is determined environmentally (Bull et al. 1988). Leopard geckos developed ovaries instead of testes when their eggs, which were incubated at temperatures that produced approximately 80% males, were injected with various doses of estradiol benzoate (Table 4). Complete sex reversal (males to females) occurred when eggs were injected with 5 µg/µL estradiol 3–6 d after laying. The results of the study indicated that leopard geckos are sensitive to gonadal feminization. A second study was conducted on leopard geckos to test the efficacy of various dosages of estradiol applied at two different stages to alter the hatchling sex ratio as well as to determine the potential use of such manipulation for conservation efforts by monitoring egg mortality and hatchling growth (Tousignant and Crews 1994). In addition to the increased mortality caused only by high dosages applied at the later stage of development, estradiol had a dose-dependent effect on the hatchling sex ratio (Table 4). Estrogen-determined females experienced normal growth and produced viable hatchlings. In a similar study, treatment of developing desert grassland whiptails, an all-female parthenogenic species of lizard, with fadrozole (a potent and specific nonsteroidal inhibitor of aromatase activity in mammals that blocks endogenous estrogen production) resulted in the inductions of male sex determination (Wibbels and Crews 1994).

B. Inorganic Contaminants

Only two studies were found that evaluated lethal and sublethal effects of inorganic contaminants on lizards (Suresh and Hiradhar 1990; Letnic and Fox 1997). Data were available for six species representing three families (Table 1). The lethal and sublethal effects of sodium fluoride on the Indian leaf-toed gecko (*Hemidactylus flaviviridis*) were studied by Suresh and Hiradhar (1990). Sodium fluoride at concentrations of 50 µg/mL enhanced the wound healing and tail regeneration of the Indian leaf-toed gecko following tail autotomy. Sodium fluoride concentrations of 100 µg/mL had no effect. Concentrations of 250, 500,

1,000, 2,000, and 2,500 µg/mL caused delayed wound healing. Geckos supplied with a sodium fluoride concentration of 3,000 µg/mL did not survive beyond the completion of wound healing. All lizards maintained on 5,000 µg/mL of sodium fluoride died in 20–30 d.

Letnic and Fox (1997) studied the responses of lizards recolonizing dry sclerophyll forest in Tomago (northwest of Newcastle in the Hunter Valley), New South Wales, Australia, following sand mining and the added impact of fluoride fallout from an aluminum smelter. In 1994, the smelter was producing 370,000 tons of aluminum/yr. Fluorides in the form of gaseous hydrogen fluoride and particulate fluorides are the principal air pollutants produced by the Tomago smelter, with approximately 760 g of fluoride being emitted/ton of aluminum produced. Six species were studied: robust ctenotus (*Ctenotus robustus*), coppertail ctenotus (*Ctenotus taeniolatus*), jacky lizard (*Amphibolurus muricatus*), forest carlia (*Carlia tetradactyla*), Guichenot's skink (*Lampropholis guichenoti*), and delicate skink (*Lampropholis delicata*). Three levels of fluoride contamination were selected: background (0.25 µg foliar fluoride/g), low (1.85–3.47 µg foliar fluoride/g), and high (5.03–12.96 µg foliar fluoride/g) in areas 3, 8, and 20 yr after sand mining. There was little evidence of fluoride fallout having a direct toxic effect. Rather, lizard species appeared to respond to changes in vegetation structure that resulted from both fluoride contamination and regeneration time since mining. The reduction in canopy cover as a result of fluoride fallout caused the absence of delicate skinks. Therefore, fluoride fallout directly influences the succession of vegetation, which indirectly retards the succession of lizard species. The additional disturbance associated with fluoride fallout appears to alter the trajectory for lizard species succession following sand mining and may, in the long term, increase the time taken for the lizard community to return to the predisturbance state. Letnic and Fox (1997) concluded that fluoride-mediated habitat change appears to be an important factor affecting vertebrate communities near sources of fluoride emissions.

C. Radionuclides/Radiation

Of the 20 studies reviewed for the effects of radionuclides/radiation, lethal effects studies were performed for eight species from six different families (Table 1). The sublethal effects of radionuclides/radiation were studied for 11 species of lizards, representing six families (Table 1). Most experiments were conducted during the 1960s, with some continuing into the 1970s (Table 5). With the exception of one study by Qazi and Jafri (1996), radionuclides/radiation effects on lizards were not studied during the 1980s and 1990s. Qazi and Jafri (1996) studied uptake and concentrations of uranium in animals and plants from a natural radioactive terrestrial ecosystem in Pakistan; however, the lizards studied were only identified as "lizards," so their study is not included here.

The activity of radioactive substances is defined as the number of nuclear disintegrations per unit time. Activity is expressed as becquerels (Bq) (1 disintegration/sec) or curies (Ci), used in this review (3.7×10^{10} disintegrations/sec).

Exposure to radiation is a measure of the amount of radiation and is defined as the change liberated per unit mass of air. We expressed exposure in röntgens (R) in this review; however, coulomb/kg (C/kg) is also used. One R is equivalent to 2.5×10^{-4} C/kg. Absorbed dose is a measure of the effect of radiation and is defined as the energy absorbed per unit mass of a substance. Its unit is the gray (Gy) or the rad, equivalent to about 1/100 Gy. In this review, absorbed dose was expressed in rads.

Eleven studies were performed on desert-dwelling lizards that were exposed continuously to ^{137}Cs from 1964 through most of the 1970s in a 9-ha, circular field enclosure in the Mojave Desert (Rock Valley) at the U.S. Department of Energy (USDOE) Nevada Test Site (Table 5). The side-blotched lizard was studied most intensely. Early studies indicated that no effects resulted from chronic low-level radiation exposure (Turner and Lannom 1968; Turner et al. 1965, 1969; French 1970; Turner 1975); however, studies performed over a longer period indicated that females and some males in all species eventually became sterile as a result of the continuous gamma radiation exposure (French 1970; Medica et al. 1973; Turner et al. 1973; Turner 1975; Turner and Medica 1977). Sterility was caused by direct radiation damage to the gonads and, in turn, gametogenesis was affected (Pearson et al. 1978). In females, ovarian tissue was absent, and the fat bodies were hypertrophied (French 1970; Medica et al. 1973; Turner et al. 1973; Turner 1975; Turner and Medica 1977). All female longnose leopard and desert horned lizards became sterile, while only some, usually older, female side-blotched lizards and western whiptails became sterile (Medica et al. 1973; Turner et al. 1973; Turner 1975; Turner and Medica 1977).

The population consequences of female sterility varied according to the life histories of the species. For example, longer-lived species with deferred sexual maturity, long life spans, and low fecundity (i.e., longnose leopard lizard, desert horned lizard) were more vulnerable than shorter-lived species with early sexual maturity and high reproductive capacity (i.e., side-blotched lizard) and eventually became extinct from the study plot (Table 5). The side-blotched lizard population could withstand the chronic stress because yearling lizards, which accounted for ~75% of the spring population, reproduced normally before damaging doses of radiation were accumulated, as most (~60%–70%) of the eggs produced were from individuals 8–9 mon old (Turner 1975). Results also indicated that the side-blotched lizard population may have been compensating for the radiation-induced sterility through a density-dependent survival mechanism (Turner 1975). The western whiptail population was also able to maintain its numbers, but the long-term prognosis was uncertain (Turner 1975).

Before the long-term study began on desert lizards in the field enclosure in Rock Valley, the effects of a detonation of a 100 (±15) kiloton device (Project Sedan) were studied at Yucca Flat, located within the USDOE Nevada Test Site (Turner and Gist 1965). The results of the study indicated that lizards were exterminated to ~1,200 m from ground zero, attributable to the physical effects of detonation (Table 5). Juveniles experienced delayed mortality as a result of habitat destruction. Turner and Gist (1965) concluded that residual gamma radiation resulting from the detonation was not likely to be lethal to the lizards.

Many natural areas are radioactive. For example, in the Upper Colorado River Basin, Utah, high surface radiation of 25–100 mR/hr occurs naturally. Radium, vanadium, and uranium have been mined from this area. Tanner (1965) studied side-blotched lizards, northern plateau lizards, and Painted Desert whiptails inhabiting four undisturbed areas with high surface radiation. There was no significant difference in external morphology between lizards that lived in areas of high surface radiation and those living in areas of low surface radiation (Tanner 1965) (Table 5).

A study was conducted in a forested, 0.25-ha, circular site in El Verde, Puerto Rico, to evaluate the effects of a 3-mon exposure to a 10,000-Ci ^{137}Cs source on two species of anoles (Turner and Gist 1970) (Table 5). Many (200–250) adult lizards were killed by a chronic dose of 84 rads/d, resulting in a zone of almost complete lethality extending 15–20 m from the source. The populations of anoles experienced some possible short-term effects such as better survival of younger lizards, better survival of large males, and altered perch height among males. In addition, because radiation caused the loss of leaves in the forest canopy, two species of anoles that are characteristic of open areas invaded the irradiated area. Considered in the context of the lizard community in the entire forest, the results were trivial (Turner and Gist 1970). The study site eventually would have been repopulated by immigrants from the surrounding forest.

A few laboratory studies have determined the effects of radiation on lizards (Smith and Thomson 1959; Brooks 1962; Dana and Tinkle 1965; Tinkle 1965; Turner et al. 1965, 1967; George and Eapen 1973; Chiplonkar and Goel 1975) (Table 5). The lethal and sublethal effects of ^{60}Co were determined for the side-blotched lizard; at a dose of 1,700–2,200 R, 50% mortality occurred within 30 d (Turner et al. 1965, 1967). At lower doses, effects on growth and survival were also observed (Table 5). The effects of X-irradiation were also determined for the side-blotched lizard in two studies (Tinkle 1965; Dana and Tinkle 1965). In one study by Tinkle (1965) with both laboratory and field components, side-blotched lizards were irradiated and then returned to study areas in West Texas. A striking decrease in lizard density the year after irradiation occurred as a result of the decrease in the number of young produced. In the second study performed in the laboratory, the $LD_{50/30}$ (within 30 d) for X-irradiation was determined to be 1,000–2,000 R, whereas sterility occurred at lower doses (Dana and Tinkle 1965) (Table 5).

Green anoles were given whole-body doses of 450–500 R/min of X-irradiation until death (Smith and Thomson 1959). The anoles died after receiving a dose of 107,000 R at 5 °C or 108,000 R at 37 °C (Smith and Thomson 1959). Brooks (1962) exposed ground skinks to doses of 300, 600, 900, 1,200, and 1,500 R at a dose rate of 99.7 R in the laboratory from a ^{60}Co source. One skink that received a dose of 600 R died after 16 d; one that received 900 R died after 16 d; and a third that received 1,500 R died after 13 d. However, because the dosages did not kill many skinks, an $LD_{50/30}$ could not be calculated. The results indicated that ground skinks are very resistant to gamma radiation.

Metabolic alterations were observed in Leschenault's leaf-toed geckos exposed to X-irradiation in a laboratory study (George and Eapen 1973). The amount of esterases in the kidneys increased while the amount of esterases decreased in the lungs; no changes were observed in the brain or liver. George and Eapen (1973) suggested that the metabolic alterations could eventually lead to pathological changes or even death. Eggs of variable agamas received either one dose of 10,250 rads, one dose of 20,500 rads, or two doses of 10,250 rads 2 d apart (total dose of 20,500 rads) in a laboratory study performed by Chiplonkar and Goel (1975). In all experiments, irradiation caused mortality, retarded development, decreased growth, hemorrhage, axial defects, unilateral microphthalmia, and eye damage in variable agama embryos.

VI. Discussion

Little information is available regarding the accumulation and effects of environmental contaminants on lizards. With the exception of those by Hall and Clark (1982), White and Krynitsky (1986), and Clark et al. (1995), all studies on the accumulation and effects of organic contaminants on lizards performed in the U.S. were conducted and published before 1980. In the U.S., concern about the effects of organochlorine pesticides was probably the greatest during the 1970s (e.g., bird eggshell thinning caused by DDT). The types of pesticides used have changed since the 1970s, but pesticide use has increased. Although DDT use was banned in the U.S. in 1972, the study performed by Clark et al. (1995) indicated that DDE, the major metabolite of DDT, was accumulating in lizards many years after the ban. Therefore, lizard studies are needed to assess the effects and accumulation of pesticides currently in use as well as the persistent organochlorine pesticides no longer used. Because most lizards eat insects, pesticides, especially insecticides, probably have the greatest overall impact on lizards relative to other environmental contaminants. For this reason, studies on the effects and accumulation of insecticides on lizards should be a top priority worldwide. Manipulative field studies and mesocosm experiments would provide the most useful information.

Studies on the effects of pesticides, or any other environmental contaminant, should consider the fact that reptile activity patterns are driven by environmental temperatures. Lizards may be protected from direct exposure from pesticide spray because of inactivity periods (Saxon 1970; Everts 1997). Many species of reptiles burrow in the soil during the hottest hours of the day, which may protect them from pesticide applications. However, this behavior increases exposure to compounds that concentrate in the subsurface layer (Everts 1997).

In the late 1980s, the U.S. Environmental Protection Agency (USEPA) proposed including the green anole in their list of species required for the preregistration testing of chemicals as part of the Federal Insecticide, Fungicide, and Rodenticide Act (FIFRA) (Hall and Henry 1992). However, anoles were never added to the list. Currently, no testing on any reptile species is required by the USEPA or any other regulatory agency for the registration or reregistration of pesticides (van der Valk 1997). An *Anolis* lizard is probably not a good predictor of responses of crocodilians and turtles, for example, even though all are in the same vertebrate class (Hall and Henry 1992). Reptiles are more diverse than any of the other terrestrial vertebrates; therefore, representative species of lizards, snakes, turtles, and crocodilians should be included in the USEPA's or any other regulatory agency's list of species on which testing should occur. However, the current trend in regulatory agencies is to develop highly standardized toxicity tests for a limited number of species, whereas the development of specialized tests for a wide variety of species is needed (van der Valk 1997).

Researchers in other countries have recognized the importance of the effects on and accumulation of pesticides in lizards, especially in desert and tropical ecosystems. In addition to some older studies, recent studies on lizards have been conducted in Australia (McIlroy et al. 1985; Birks and Olson 1987; Twigg et al. 1988; Twigg and Mead 1990), Africa (Lambert 1997a), Turkey (Özelmas and Akay 1995), and Spain (Fossi et al. 1995; Sanchez et al. 1997). These results could be used for ecological risk assessments performed in the U.S., as well as for those performed in other countries, especially for pesticides that are no longer used but are persistent in the environment. Studies by Lambert (1993), Fossi et al. (1995), and Sanchez et al. (1997) indicate that lizards are ideal bioindicators for pesticide exposure.

The effects of environmental estrogens or endocrine-disrupting contaminants have recently begun to receive attention. Crain and Guillette (1997) reviewed the effects of endocrine-disrupting contaminants on reproduction in vertebrate wildlife. To our knowledge, the effects of endocrine-disrupting contaminants have not been studied in lizards. However, the studies by Bull et al. (1988), Tousignant and Crews (1994), and Wibbels and Crews (1994) included in this review indicate that lizards would be susceptible to effects of endocrine-disrupting contaminants because experiments on lizards with estradiol and fadrozole resulted in sex reversal, in addition to other effects (Table 4). Exposure of eggs of the red-eared turtle (*Trachemys scripta elegans*) to polychlorinated biphenyls (PCBs) caused feminization, even though the eggs were incubated at male-producing temperatures (Bergeron et al. 1994; Crews et al. 1995). Some lizard species are similar to alligators and turtles in which sex determination is dependent on egg incubation temperature.

Contamination from heavy metals resulting from such sources as mining, industrial emissions, and the operation of motor vehicles is a serious problem recognized in most countries (Loumbourdis 1997). However, very little information was available on the bioaccumulation of heavy metals in lizards. Only four studies were found that contained data: one in Brazil (Schmidt 1984), one in India (Kaur 1988), one in Puerto Rico (Burger et al. 1992), and one in Greece (Loumbourdis 1997). To our knowledge, no studies of heavy metal residues in lizards have been performed in the U.S. Because lizards have been shown to accumulate heavy metals (Schmidt 1981, 1984; Kaur 1988; Loumbourdis 1997) and serve as prey items for many species of wildlife (especially birds), whole-body heavy metal concentration data would be useful for those performing ecological risk assessments. The studies by Schmidt (1981, 1984), Kaur (1988), and Loumbourdis (1997) indicated that lizards are also good bioindicators for heavy metal contamination. Only two additional studies on the lethal and sublethal effects of metals were found. Studies of the lethal and sublethal effects and bioaccumulation of inorganic contaminants are seriously lacking.

Studies on radiation effects are important because many sites throughout the world are contaminated with radioactive waste. For example, many contaminated sites in the U.S. have resulted from current and earlier USDOE operations. Data available for this review were collected during the 1960s and 1970s, and up-to-date information is needed. Lizards have relatively small home ranges and could be receiving continuous exposure in contaminated areas. However, as observed in the studies performed at the Nevada Test Site, the age of sexual maturity of the species ultimately determined the effect of female sterility on the population.

Field studies performed at the Nevada Test Site (Turner 1965) and in Puerto Rico (Turner and Gist 1970) provide useful data that could be used in terrestrial ecological risk assessments which include lizards. A limitation to the studies is that populations were exposed only to external, continuous gamma radiation. Lizards would also be exposed internally to radionuclides inhaled or assimilated at waste disposal sites or areas contaminated by a spill or accident. Many current and proposed radioactive waste disposal sites are or will be located in desert ecosystems, which also contain a high diversity of lizards. Studies are lacking on the effects of lizards living in areas contaminated with radioactive waste. Information on whole-body concentrations of accumulated radionuclides in lizards is also needed because lizards serve as a food source for many higher vertebrates.

The largest number of studies reviewed were for species in the families Phrynosomatidae, Teiidae, and Scincidae (Table 1). No studies were found for lizards in the families Opluridae, Tropiduridae, Hoplocercidae, Corytophanidae, Pygopodidae, Dibamidae, Gymnophthalmidae, Xantusiidae, Cordylidae, Gerrhosauridae, Xenosauridae, and Lanthanotidae. Most of these families contain few species and have limited distributions (Pough et al. 1998), so the lack of information was not surprising. Characteristics of these families suggest that the species within them may already be rare. Therefore, studies on species included in these families would be important if any are at risk of exposure to environmental contaminants. The lack of research on Tropiduridae and Gymnophthalmidae is of concern because both families are fairly large. The family Tropiduridae contains approximately 270 species and is distributed in South America, the West Indies, and the Galápagos Islands, while Gymnophthalmidae includes approximately 140 species and has a geographic distribution from southern Mexico to Argentina (Pough et al. 1998). As both families are mainly distributed in South America, it could be that studies on these lizards were published in more obscure Latin American journals and did not appear in literature or database searches.

No studies were found on the effects or accumulation of environmental contaminants on lizard species currently listed as threatened or endangered. Studies conducted on closely-related lizard species could be used for ecological risk assessments on threatened and endangered species. For example, the green anole has been called the "green mouse" by herpetologists because it has been studied so extensively (A.C. Echternacht, personal communication); unfortunately, few studies have examined the effects of environmental contaminants on the green anole. However, caution must be taken in extrapolating data from one species to another. As was shown in the long-term studies at the Nevada Test Site (Turner 1975), the effects of sterility resulting from chronic low-level exposure to radiation had different results in side-blotched and longnose leopard lizards, species from the same family, because of the differences in their life histories.

This review focused only on concentrations in and the effects of environmental contaminants on lizards. However, other stressors, such as habitat loss or introduction of exotic species, could be the focus of ecological risk assessments. Also, lizards may be exposed to more than one stressor or mixtures of environmental contaminants. For example, in an agricultural area lizards could be exposed to mixtures of pesticides. In addition, heavy metal contamination is a result of combinations of heavy metals originating from motor vehicles, industrial emissions, and urban runoff. Furthermore, radioactive waste usually consists of more than one type of radionuclide. Lizards may also be affected by contaminants in combination with some type of environmental disturbance. Only one study explored effects of multiple stressors (Lentic and Fox 1997). Effects of multiple stressors have recently begun to receive attention. Lastly, no lizard studies were found regarding the use of chemical residues in tissues as a basis for linkage of exposure to effects for ecological risk assessment.

Summary

This review is the most comprehensive currently available of the effects and accumulation of environmental contaminants on lizards. The importance of lizards was emphasized in hope that they be included in ecological risk assessments as well as any study on environmental contamination. Some studies presented here indicated that lizards are ideal bioindicators. They are important as a component of biodiversity, and many species are listed as threatened or endangered. In addition, lizards are a significant part of many ecosystems as well as an important link in many food chains. There are large gaps in data for many environmental contaminants, particularly data on lizards. Ecotoxicological studies on a wide variety of lizard species are needed; both laboratory and field studies would provide useful information. Because the majority of lizards are insectivores, studies of the effects and accumulation of pesticides are essential. A few current studies are available from Africa and Australia, but most, especially those conducted in the U.S., were not current. Studies are needed on the effects and accumulation of ubiquitous contaminants, such as heavy metals and PCBs. Because of the many contaminated sites and the significant waste disposal problem, studies are needed on the effects and accumulation of radionuclides on lizards. Furthermore, effects of multiple stressors must be studied. Last, studies are needed linking the effects of contaminants to tissue residues. It is hoped that the convenience of having the vast majority of lizard environmental contaminant data available in one document will encourage its use.

Acknowledgments

We thank Cristof Stumpf for English translations from German and Kenichi Miyamoto for English translations from Japanese.

References

Beck AB (1956) The copper content of the liver and blood of some vertebrates. Aust J Zool 4:1–18.
Bergeron JM, Crews D, McLachlan JA (1994) PCBs as environmental estrogens: turtle sex determination as a biomarker of environmental contamination. Environ Health Perspect 102:780–781.
Best SM (1973) Some organochlorine pesticide residues in wildlife of the Northern Territory, Australia, 1970–71. Aust J Biol Sci 26:1161–1170.
Birks PR, Olsen AM (1987) Pesticide concentrations in some south Australian birds and other fauna. Trans R Soc S Aust 111:67–77.
Bishop CA, Gendron AD (1998) Reptiles and amphibians: shy and sensitive vertebrates of the Great Lakes Basin and St. Lawrence River. Environ Monit Assess 53:225–244.
Brooks GR (1962) Resistance of the ground skink *Lygosoma laterale* to gamma radiation. Herpetologica 18(2):128–129.
Bryan AM, Stone WB, Olafsson PG (1987) Disposition of toxic PCB congeners in snapping turtle eggs: expressed as toxic equivalents of TCDD. Bull Environ Contam Toxicol 39:791–796.
Bull JJ, Gutzke WHN, Crews D (1988) Sex reversal in estradiol in three reptilian orders. Gen Comp Endrocrinol 70:425–428.
Burger J, Cooper K, Saliva J, Gochfield D, Lipsky D, Gochfield M (1992) Mercury bioaccumulation in organisms from three Puerto Rican estuaries. Environ Monit Assess 22:181–197.
Chiplonkar JM, Goel SC (1975) Effects of γ-rays on the developing embryos of *Calotes versicolor*. Experientia (Basel) 31:1213–1215.
Clark DR Jr, Flickinger EL, White DH, Hothem RL, Belisle AA (1995) Dicofol and DDT residues in lizard carcasses and bird eggs from Texas, Florida, and California. Bull Environ Contam Toxicol 54:817–824.
Conant R, Collins JT (1991) A Field Guide to Reptiles and Amphibians of Eastern and Central North America. Houghton Mifflin, Boston, MA.
Crain DA, Guillette LJ Jr (1997) Endocrine-disrupting contaminants and reproduction in vertebrate wildlife. Rev Toxicol 1:47–70.
Crain DA, Guillette LJ Jr (1998) Reptiles as models of contaminant-induced endocrine disruption. Anim Reprod Sci 53:77–86.

Crews D, Bergeron JM, McLachlan JA (1995) The role of estrogen in turtle sex determination and the effect of PCBs. Environ Health Perspect 103(suppl 7):73–77.

Culley DD, Applegate HG (1967a) Insecticide concentrations in wildlife at Presidio, Texas. Pestic Monit J 1(2):21–28.

Culley DD Jr, Applegate HG (1967b) Pesticides at Presidio. IV. Reptiles, birds, and mammals. Tex J Sci 19:301–310.

Dana SW, Tinkle DW (1965) Effects of X-radiation on the testes of the lizard, *Uta stansburiana stejnegeri*. Int J Radiat Biol 9:67–80.

Everts JW (1997) Ecotoxicology for risk assessment in arid zones: some key issues. Arch Environ Contam Toxicol 32:1–10.

Fontenot LW, Noblet GP, Platt SG (1996) A survey of herpetofauna inhabiting polychlorinated biphenyl contaminated and reference watersheds in Pickens County, South Carolina. J Elisha Mitchell Sci Soc 112(1):20–30.

Fossi MC, Sanchez-Hernandez JC, Diaz-Diaz R, Lari L, Garcia-Hernandez JE, Gaggi C (1995) The lizard *Gallotia galloti* as a bioindicator of organophosphorus contamination in the Canary Islands. Environ Pollut 87:289–294.

Frank N, Ramus E (1995) A Complete Guide to Scientific and Common Names of Reptiles and Amphibians of the World. NG Publishing, Pottsville, PA.

French NR (1970) Chronic low-level gamma irradiation of a desert ecosystem for five years. In: Grauby A (ed) Actes du Symposium International de Radioecologie, Centre d'Etudes Nucleaires de Cadarache, Cadarache, France, pp 1151–1165.

George KC, Eapen J (1973) Effect of X-irradiation on esterases of tissues of house lizard, *Hemidactylus leschenaulti* Dum. & Pipr. Indian J Exp Biol 11:76–78.

Hall RJ (1980) Effects of environmental contaminants on reptiles: a review. Special Scientific Report—Wildlife No. 228. U.S. Department of the Interior, Fish and Wildlife Service, Washington, DC.

Hall RJ, Clark DR Jr (1982) Responses of the iguanid lizard *Anolis carolinensis* to four organophosphorus pesticides. Environ Pollut Ser A 28:45–52.

Hall RJ, Henry PFP (1992) Assessing effects of pesticides on amphibians and reptiles: status and needs. Herpetol J 2:65–71.

Kaur S (1988) Lead in the scales and cobras and wall lizards from rural and urban areas of Punjab, India. Sci Total Environ 77(2-3):289–290.

Kihara H, Yamashita H (1978) Lethal effects of various agricultural chemicals against reptiles (in Japanese). Snake 10:10–15.

Lambert MRK (1993) Effects of DDT ground-spraying against tsetse flies on lizards in NW Zimbabwe. Environ Pollut 82:231–237.

Lambert MRK (1997a) Effects of pesticides on amphibians and reptiles in sub-Saharan Africa. Rev Environ Contam Toxicol 150:31–73.

Lambert MRK (1997b) Environmental effects of heavy spillage from a destroyed pesticide store near Hargeisa (Somaliland) assessed during the dry season, using reptiles and amphibians as bioindicators. Arch Environ Contam Toxicol 32:80–93.

Lance VA, Cort T, Masuoka J, Lawson R, Saltman P (1995) Unusually high zinc concentrations in snake plasma, with observations on plasma zinc concentrations in lizards, turtles and alligators. J Zool Lond 235:577–585.

Lentic MI, Fox BJ (1997) The impact of industrial fluoride fallout on faunal succession following sand mining of dry sclerophyll forest at Tomago, NSW. I. Lizard recolonisation. Biol Conserv 80:63–81.

Loumbourdis NS (1997) Heavy metal contamination in a lizard, *Agama stellio stellio*, compared in urban, high altitude and agricultural, low altitude areas of north Greece. Bull Environ Contam Toxicol 58:945–952.

McIlroy JC, King DR, Oliver AJ (1985) The sensitivity of Australian animals to 1080 poison. VIII. Amphibians and reptiles. Aust Wildl Res 12:113–118.

Medica PA, Turner FB, Smith DD (1973) Effects of radiation on a fenced population of horned lizards. J Herpetol 7:79–85.

Meyers-Schöne L, Walton BT (1994) Turtles as monitors of chemical contaminants in the environment. Rev Environ Contam Toxicol 135:93–153.

Meyers-Schöne L, Shugart LR, Beauchamp JJ, Walton BT (1993) Comparison of two freshwater turtle species as monitors of radionuclide and chemical contamination: DNA damage and residue analysis. Environ Toxicol Chem 12:1487–1496.

Nagy KA, Medica PA (1985) Altered energy metabolism in an irradiated population of lizards at the Nevada Test Site. Radiat Res 103:98–104.

Olafsson PG, Bryan AM, Bush B, Stone W (1983) Snapping turtles—a biological screen for PCBs. Chemosphere 12:1525–1532.

Overmann SR, Krajicek JJ (1995) Snapping turtles (*Chelydra serpentina*) as biomonitors of lead contamination of the Big River in Missouri's Old Lead Belt. Environ Toxicol Chem 14:689–695.

Özelmas U, Akay MT (1995) Histopathological investigations of the effects of malathion on dwarf lizards (*Lacerta parva*, Boulenger 1887). Bull Environ Contam Toxicol 55: 730–737.

Pearson AK, Licht P, Nagy KA, Medica PA (1978) Endocrine function and reproductive impairment in an irradiated population of the lizard *Uta stansburiana*. Radiat Res 76: 610–623.

Pianka ER (1986) Ecology and Natural History of Desert Lizards. Princeton University Press, Princeton, NJ.

Pough FH, Andrews RM, Cadle JE, Crump ML, Savitzky AH, Wells KD (1998) Herpetology. Prentice-Hall, Upper Saddle River, NJ.

Qazi JI, Jafri RH (1996) Uptake and concentration of uranium in animals and plants from a natural radioactive terrestrial ecosystem in Pakistan. Punjab Univ J Zool 11: 51–56.

Rodda GH, Perry G, Rondeau RJ (1999) The densest terrestrial vertebrate. In: Program Book and Abstracts of the Joint Meeting of the American Society of Ichthyologists and Herpetologists, American Elasmobranch Society, Herpetologists' League, and Society for the Study of Amphibians and Reptiles, Pennsylvania State University, State College, PA, June 24–30, 1999, p 195.

Roughgarden J (1995) Anolis Lizards of the Caribbean: Ecology, Evolution, and Plate Tectonics. Oxford Series in Ecology and Evolution. Oxford University Press, New York, NY.

Sanchez JC, Fossi MC, Focardi S (1997) Serum B esterases as a nondestructive biomarker in the lizard *Gallotia galloti* experimentally treated with parathion. Environ Toxicol Chem 16:1954–1961.

Saxon JG (1970) The biology of the lizard, *Cnemidophorus tesselatus*, and effects of pesticides upon the population in the Presidio Basin, Texas. Ph.D. Dissertation, Texas A & M University, College Station, TX.

Schmidt J (1981) Lead and cadmium residues by inner and outer city *Lacerta* populations (in German). Verh Ges Ökol (Berl) 9:297–300.

Schmidt J (1984) Heavy metal analyses in *Hemidactylus mabouia* (Geckonidae) as a

method to classify urban environmental quality (in German). Amazoniana 9(1): 35–42.
Schwartz A, Henderson RW (1991) Amphibians and Reptiles of the West Indies: Descriptions, Distributions, and Natural History. University of Florida Press, Gainesville, FL.
Smith DE, Thomson JF (1959) Physiological and biochemical studies on various species exposed to massive X-irradiation. Radiat Res 11:198–205.
Stebbins RC (1985) A Field Guide to Western Amphibians and Reptiles. Houghton Mifflin, Boston, MA.
Suresh B, Hiradhar PK (1990) Toxicity of NaF on tail regeneration in gekkonid lizard *Hemidactylus flaviviridis*. Indian J Exp Biol 28:1086–1087.
Tanner WW (1965) A comparative population study of small vertebrates in the uranium areas of the Upper Colorado River Basin of Utah. Brigham Young University Science Bulletin, Biological Series—Vol. VII, No. 1. Brighman Young University, UT.
Tinkle DW (1965) Effects of radiation on the natality, density and breeding structure of a natural population of lizards, *Uta stansburiana*. Health Phys 11:1595–1599.
Tousignant A, Crews D (1994) Effects of exogenous estradiol applied at different embryonic stages on sex determination, growth, and mortality in the leopard gecko (*Eublepharis macularius*). J Exp Zool 268:17–21.
Turner FB (1975) Effects of continuous irradiation on animal populations. Adv Radiat Biol 5:83–144.
Turner FB, Gist CS (1965) Influences of a thermonuclear cratering test on close-in populations of lizards. Ecology 46:845–852.
Turner FB, CS Gist (1970) Observations of lizards and tree frogs in an irradiated Puerto Rican forest. In: Odum HT, Pigeon RF (eds) A tropical rain forest, a study of irradiation and ecology at El Verde, Puerto Rico. TID-24270. U.S. Atomic Energy Commission, Division of Technical Information, Oak Ridge, TN, pp E-25–E-49.
Turner FB, Lannom JR Jr (1968) Radiation doses sustained by lizards in a continuously irradiated natural enclosure. Ecology 49:548–551.
Turner FB, Medica PA (1977) Sterility among female lizards (*Uta stansburiana*) exposed to continuous γ irradiation. Radiat Res 70:154–163.
Turner FB, Hoddenbach GA, Lannom JR Jr (1965) Growth of lizards in natural populations exposed to gamma irradiation. Health Phys 11:1585–1593.
Turner FB, Lannom JR, Kania HJ, Kowalewsky BW (1967) Acute gamma irradiation experiments with the lizard *Uta stansburiana*. Radiat Res 31:27–35.
Turner FB, Medica PA, Lannom JR Jr, Hoddenbach GA (1969) A demographic analysis of continuously irradiated and nonirradiated populations of the lizard, *Uta stansburiana*. Radiat Res 38:349–356.
Turner FB, Licht P, Thrasher JD, Medica PA, Lannom JR Jr (1973) Radiation-induced sterility in natural populations of lizards (*Crotophytus wislizenii* and *Cnemidophorus tigris*). In: Nelson DJ (ed) Radionuclides in Ecosystems. Proceedings of the Third National Symposium on Radioecology, May 10–12, 1971, Oak Ridge, TN, pp 1131–1143.
Twigg LE, Mead RJ (1990) Comparative metabolism of, and sensitivity to, fluoroacetate in geographically separated populations of *Tiliqua rugosa* (Gray) (Scincidae). Aust J Zool 37:617–626.
Twigg LE, King DR, Bradley AJ (1988) The effect of sodium monofluoroacetate on plasma testosterone concentration in *Tiliqua rugosa* (Gray). Comp Biochem Physiol 91C:343–347.

van der Valk HCHG (1997) Community structure and dynamics in desert ecosystems: potential implications for insecticide risk assessment. Arch Environ Contam Toxicol 32:11–21.

Walker CH, Ronis MJJ (1989) The monooxygenases of birds, reptiles and amphibians. Xenobiotica 19(10):1111–1121.

Wheeler WB, Jouvenaz DP, Wojcik DP, Banks WA, Van Middelem CH, Lofgren CS, Nesbitt S, Williams L, Brown R (1977) Mirex residues in nontarget organisms after application of 10-5 bait for fire ant control, Northeast Florida—1972-74. Pestic Monit J 11:146–156.

White DH, Krynitsky AJ (1986) Wildlife in some areas of New Mexico and Texas accumulate elevated DDE residues, 1983. Arch Environ Contam Toxicol 15:149–157.

Wibbels T, Crews D (1994) Putative aromatase inhibitor induces male sex determination in a female unisexual lizard and in a turtle with temperature-dependent sex determination. J Endocrinol 141:295–299.

Wojcik DP, Banks WA, Wheeler WB, Jouvenaz DP, Van Middelem CH, Lofgren CS (1975) Mirex residues in nontarget organisms after application of experimental baits for fire ant control, Southwest Georgia—1971-1972. Pestic Monit J 9:124–133.

Wright JW, Vitt LJ (1993) Biology of Whiptail Lizards (Genus *Cnemidophorus*). Oklahoma Museum of Natural History, Norman, OK.

Manuscript received May 10, 1999; accepted August 20, 1999.

Biomarkers in Earthworms

Janeck J. Scott-Fordsmand and Jason M. Weeks

Contents

I. Introduction	117
II. DNA Alterations	118
III. Metallothionein and Other Metal-Binding Proteins	120
IV. Cholinesterases	122
V. Other Enzymatic Responses	126
A. Cytochrome P-450/Mixed-Function Oxidase (MFO)	126
B. Glutathione S-Transferase (GST)	126
C. Aminolevulinic Acid Dehydratase (ALAD)	131
D. Peroxidases	131
E. Catalases	131
F. Others	132
VI. Energy Reserve Responses	132
VII. Lysosomal Membrane Integrity	133
VIII. Immunological Responses	135
IX. Sperm Quality Responses	138
X. Neurological Responses	139
XI. Histopathological Responses	141
XII. Behavioral Responses	144
XIII. Discussion	146
Summary	147
Acknowledgment	148
References	148

I. Introduction

Earthworms are abundant and very important organisms for both soil formation and the breakdown of organic matter in most terrestrial ecosystems. They are, in many ecosystems, key species responsible for moving large amounts of soil from deeper horizons to the surface layers (Edwards and Bohlen 1992). For these and other reasons, earthworms have been used extensively in ecotoxicological soil studies (EEC 1985; Kokta 1992; OECD 1984; van Gestel et al. 1989a,b, 1995).

Communicated by Dr. Finn Bro-Rasmussen
J.J. Scott-Fordsmand
National Environmental Research Institute, Department of Terrestrial Ecology, Vejlsøvej 25, P.O. Box 314, DK-8600 Silkeborg, Denmark (✉)
J.M. Weeks
Institute of Terrestrial Ecology, Monks Wood, Abbots Ripton, Huntingdon, Cambridgeshire PE17 2LS, UK

More recently, the uses of biomarkers to estimate either exposure or resultant effects of chemicals have received considerable attention (Fossi and Leonzio 1994; Hugget et al. 1992; McCarthy and Shugart 1990). Biomarkers has been defined as "a biological response that can be related to an exposure to, or toxic effect of, an environmental chemical and chemicals" (Peakall and Shugart 1993). The major reasons for the current interest in biomarkers are the limitations of the classical approach to environmental toxicology. In the classical approach, the chemical concentration present in the ambient environment is related to adverse organism effects such as mortality, reproduction, or growth. However, under various environmental conditions the availability of the toxicant may differ, rendering it difficult to extrapolate laboratory toxicity data to variable field conditions. Biomarkers address the question of toxicity by only quantifying the bioactive fraction of the pollutants. They may also integrate exposure or effects of complex pollution and may be applicable under both laboratory and field conditions (Weeks 1996).

For a biomarker, or a battery of such, to be used in risk assessment some basic understanding is needed. The most basic requirement is that the biomarker can be measured in an organism relevant for the ecosystem investigated. Knowledge is required about the range of hazards that elicit a biomarker response. To estimate the magnitude of the problem, a dose–response relationship between the exposure level and a biomarker response is essential. It should be mentioned that this need not be straightforward as the biomarker response reflects only the bioactive fraction, which may or may not increase or decrease in the same way as the external dose of the pollutant. Biomarkers may function as early warning indicators at higher levels of organization if there is a correlation (link) between the observed biomarker response and deleterious changes to the organisms or its reproductive output. To reflect *in situ* pollution, the biomarker response should also ideally have a low inherent variability with a low, or at least a known, dependence on physiological and physiochemical conditions. The induction time, and the persistence of the biomarker response, should be known to estimate the likelihood and significance of detecting a response in field samples.

In this review, we attempt to draw together current knowledge on potential and known biomarkers presently employed in earthworms. We describe the key features investigated for these markers and discuss the development of such as useful tools for legislative purposes.

II. DNA Alterations

In earthworms, chemical compounds or their metabolites may have genotoxic properties. Chemicals interfering with the DNA may lead to covalent binding of the chemical or its metabolites forming DNA adducts, strand breakage of the DNA, base exchange in the DNA, or increased unscheduled DNA synthesis (Fairbairn et al. 1995; Lloyd-Jones 1995; Shugart et al. 1992; van Welie et al. 1992). These changes may severely affect the cell and hence the organism.

The effects of chemical exposure upon earthworm DNA have been tested in only three species, *Lumbricus terrestris*, *L. castaneus*, and *Eisenia fetida* (Table 1). Strand breakages have been shown for a variety of chemicals. For example, Verschaeve et al. (1993) observed DNA single-strand breakage in coelomocytes of *L. terrestris* exposed to dioxin-contaminated soil. Strand breakage was also found following exposure to mitomycin C, coke pollution, and soils containing a mixture of benzene, aniline, and other organic compounds (Salagovic et al. 1996; Verschaeve and Gilles 1995). Chemicals may also cause DNA adducts. For example, Walsh et al. (1995) and Van Schooten et al. (1995) recorded an increased rate in the production and frequency of DNA adducts, using a ^{32}P-postlabeling technique, in *L. terrestris* exposed to PAH-contaminated soils. A dose–response relationship between pollutant and strand breakage frequency has been shown for worms exposed to dioxin, mitomycin C, and coke (Salagovic et al. 1996; Verschaeve and Gilles 1995; Verschaeve et al. 1993). Such a relationship has not been observed for DNA adduct induction.

No data were found linking DNA changes in earthworms to higher-level effects. Such a relationship may be complicated if DNA alterations are repaired by DNA-repair mechanisms or if damage is restricted to inactive site(s) of the DNA and hence is not important for the health of the organism. However, if the damage is not repaired and is located at an active site, a severe effect on the parent or the following generations may result (Anderson and Wild 1994).

The natural variation of strand breakage and DNA adducts in earthworms has not been described. However, studies on other organisms have shown that strand breakage and DNA adducts occur under normal conditions, and it is therefore important to have appropriate reference material when studying such effects (Lloyd-Jones 1995; Shugart and Theodorakis 1994). Little is known about the persistence of these markers, but it has been shown that DNA adducts

Table 1. DNA Alterations in Earthworms in Response to Toxic Compounds.

DNA alterations	Chemical	Species[a]	Reference
Strand breaks	Cadmium (Cd), pentachlorophenol (PCP), phenol, trifluralin	*Eisenia fetida*	Bierkens et al. (1998)
Strand breaks	Coke-polluted soil	*E. fetida*	Salagovic et al. (1996)
Adducts	Polynuclear aromatic hydrocarbons (PAH)	*Lumbricus terrestris*	van Schooten et al. (1995)
Strand breaks	X-rays, mitomycin	*L. terrestris, E. fetida*	Verschaeve and Gilles (1995)
Strand breaks	Dioxin	*L. castaneus*	Verschaeve et al. (1993)
Adducts	PAH	*L. terrestris*	Walsh et al. (1995)
Adducts	PAH	*E. fetida*	Walsh et al. (1997)

[a]Species names in all tables according to Sims and Gerard (1985).

measured following exposure to PAH may persist for some considerable time and even increase with duration of exposure. In contrast, strand breakages may be repaired rather rapidly (van Schooten et al. 1995; Walsh et al. 1995).

III. Metallothionein and Other Metal-Binding Proteins

A large number of studies have shown the ability of heavy metal-exposed earthworms to accumulate certain metals (Hopkin 1989). Heavy metals entering earthworms may be bound in various ways within the organism. One way of binding, and detoxifying, heavy metals in tissues is by binding these metals to metallothionein (MT) or other metal-binding proteins (MBP). Metallothionein and other MBPs are known to occur throughout the animal kingdom, as well as in several plants and eukaryotic microorganisms, but are little studied in terrestrial invertebrates (Dallinger 1996; Stegeman et al. 1992). In invertebrates, MBPs may share characteristics with vertebrate MTs, but the composition of the invertebrate MTs differs among species (Dallinger 1996).

Metal-binding proteins have been identified in *Eisenia fetida*, *Dendrodrilus rubidus*, *Aporrectodea caliginosa*, and various *Lumbricus* species (Table 2). These MBPs in some cases resembled the MT found in other vertebrates, although in general they contained less cysteine, but in other cases they were quite distinct from mammalian MTs and contained large quantities of aromatic amino acids (Bauer-Hilty et al. 1989; Dallinger 1996; Morgan et al. 1989; Nejmeddine et al. 1992; Stürzenbaum et al. 1998a; Yamamura et al. 1981).

A number of metals and some organic compounds are known to induce MTs and MBPs (Dallinger 1996; Roesijadi 1993). In earthworms, MBP induction has been investigated almost entirely in relation to cadmium exposure. For example, Suzuki et al. (1980) observed an induction of three cadmium-binding proteins in *E. fetida* following cadmium exposure. One of the three cadmium-binding proteins was characterized as a MT, although the amino acid composition was somewhat different from MT identified in rabbits (Yamamura et al. 1981). Using cadmium-exposed *Aporrectodea caliginosa*, Nejmeddine et al. (1992) isolated a cadmium-binding protein that was different from MTs described in mammals and other invertebrates.

In *Dendrobaena octaedra*, two cadmium-binding proteins could be isolated that resembled MTs in *E. fetida*. A Zn-binding molecule similar to those observed in mammals was also found in the worms (Bengtsson et al. 1992; Yamamura et al. 1981).

For *Lumbricus* species, Bauer-Hilty et al. (1989) observed an induction of a low molecular weight cadmium-binding protein in *L. variegans* following cadmium exposure. The protein resembled MBPs from snails but was different from rabbit MT. Furst and Nguyen (1989) and Ramseier et al. (1990) found cadmium-bound MT-like proteins similar to those observed in rodents. Morgan et al. (1989) isolated two cadmium-binding proteins in *L. rubellus* and one in *Dendrodrilus rubidus* following exposure to waste from a Cd-Zn mine area. All resembled MT by having low or no aromatic amino acid or histidine content.

Table 2. Metallothionein (MT) and Other Metal-Binding Proteins (MBP) in Earthworms in Response to Toxic Compounds.

MT and MBP	Chemical	Species	Reference
Cd-binding protein	Cd	*Lumbricus variegatus*	Bauer-Hilty et al. (1989)
Cu-, Zn-, and Cd-binding proteins	Copper (Cu), Zinc (Zn), Cd	*Dendrobaena octaedra*	Bengtsson et al. (1992)
Cd-binding protein	Cd	*Eisenia fetida*	Dallinger (1994)
Cd-binding protein	Cd	*L. terrestris*	Furst and Nguyen (1989)
Histidine	Cu	*L. rubellus, E. andrei*	Gibb et al. (1997)
Cd-binding protein	Cd	*E. fetida*	Grüber et al. (1998)
Cd-binding protein	Cd	*L. rubellus*	Mariño et al. (1998)
Cd-binding protein	Lead (Pb)/Zn mine spoil	*L. rubellus, Dendrodilus rubidus*	Morgan et al. (1989)
Cd-binding protein	Cd	*Aporrectodea caliginosa*	Nejmeddine et al. (1992)
Cd-binding protein	Cd	*L. terrestris*	Ramseier et al. (1990)
Gene expression amplification	Cd	*L. rubellus*	Stürzenbaum et al. (1996)
Gene expression amplification	Cd	*L. rubellus*	Stürzenbaum et al. (1998a)
Cd-binding protein	Cd	*L. rubellus*	Stürzenbaum et al. (1998b)
Gene expression amplification	Cd	*L. rubellus*	Stürzenbaum et al. (1998c)
Cd-binding protein	Cd	*E. fetida*	Suzuki et al. (1980)
Cd-binding protein	Cd sludge	*E. fetida*	Yamamura et al. (1981)

Recently, Mariño et al. (1998) observed a MT homolog binding Cu and Cd, but not Zn, in *L. rubellus*. Stürzenbaum et al. (1998a) have isolated and sequenced the earthworm MT from *L. rubellus* exposed to Cd-contaminated soil. The MT had a high cysteine content and no significant aromatic residues. Two isoforms were separated. They have further purified a gene responsible for MT production (Stürtzenbaum et al. 1998b,c). After exposing *L. rubellus* to raised soil copper concentrations Gibb et al. (1997) found a dose-dependent increase in histidine after copper exposure. It was further revealed that the internal copper was associated with this amino acid (Gibb et al. 1997).

A number of studies have shown the induction of cadmium MT-like proteins and other Cd MBP in a dose–response manner following cadmium exposure of earthworms. For example, in *E. fetida* a dose–response-related induction of three cadmium-binding proteins was observed following cadmium exposure (Suzuki et al. 1980).

It is unknown whether induction of MBPs relates to effects at higher organizational levels, such as reduced reproduction of the earthworms. Given the fact that the roles of MTs and MBPs are far from understood and may be related to internal metal transfer or storage, it is not surprising that such linkages have yet to be established. It seems that these proteins may in the first instance be considered as markers of exposure to chemical stressors, notably cadmium (Stegeman et al. 1992).

There have been no studies of the influence of natural stressors or time on MTs or MBPs in earthworms. In other organisms, natural stressors (e.g., temperature, nutritional status, hormone level, and organic compounds) have been observed to influence MT and MBP induction (Dallinger 1996; Engle and Brouwer 1987; Min et al. 1993; Roesijadi 1993). Metal-binding proteins may also be metabolized in certain tissues, as observed in certain aquatic invertebrates and vertebrates (Bremner et al. 1978; Klassen et al. 1993; Riddlington et al. 1981; Stegeman et al. 1992).

IV. Cholinesterases

Earthworms use cholinesterases (ChE), as do many other organisms, in the nervous system (Stenersen 1980a,b; Teräväinen 1969; Vigh-Teichmann and Goslar 1968). Few studies have characterized ChEs in earthworms. Andersen et al. (1978) reported only one ChE in *Eisenia fetida*, but Stenersen (1979c, 1980a,b) reported two ChEs in the same species. One (E1) of the esterases was regarded as propionylcholinesterase and related to acetylcholinesterase (AChE); the other (E2) was regarded as nonspecific procholinesterase, but with several properties different from ChE of vertebrates (Stenersen 1979c, 1980a,b).

In earthworms, the inhibition of ChEs by chemical compounds has been investigated in *Lumbricus* species (*L. terrestris* and *L. rubellus*), *Eisenia* species (*E. fetida, E. andrei,* and *E veneta*), *Allolobophora chlorotica, Aporrectodea caliginosa, Lampito mauritii,* and *Pheretima posthuma* (Table 3).

A wide variety of compounds have been tested for potential inhibition of the ChE activity of earthworms, including organophosphorous (OPs), carbamates, organochlorines, benzimidazole, and inorganic compounds (Table 3). In vertebrates it has long been established that two classes of chemical compounds, the carbamate and OP pesticides, can cause a depression in total ChE activity (Edwards and Fischer 1991).

OPs in general give the most severe reduction of earthworm ChE activity, with less overall reduction following exposure to carbamates (Stenersen et al.

Table 3. Cholinesterase (ChE) Activity in Earthworms in Response to Toxic Compounds.

Chemical[a]	Species	Reference
Phosphamidon (OP), monocrotophos (OP), dichlorvos	*Lampito mauritii*	Bharathi and Rao (1985)
Carbaryl (Car)	*Pheretima posthuma*	Dikshith and Gupta (1981)
Methiocarb (Car)	*Lumbricus terrestris*	Grafton (1995)
Carbaryl (Car)	*P. posthuma*	Gupta and Sundaraman (1991)
Pb, Uranium (U)	*Eisenia fetida*	Labrot et al. (1996)
Azinphos-methyl (OP), benomyl (Benz), carbendazim (Benz), copper oxychloride, endosulfan (OC), methidathion (OP), methomyl (Oxime Car), parathion-ethyl (OP), propoxur (Car), thiabendazol (Benz), thiophanate methyl (Benz)	*L. terrestris*	Niklas (1979)
Cd	*E. fetida*	Scaps et al. (1997)
Pb		
Aldicarb (Car), carbaryl (Car), carbofuran (Car), oxamyl (Oxime Car), paraoxon (OP), parathion (OP), gamma-HCH (OC), dithiocarbamate (CAR)	*E. fetida, Aporrectodea caliginosa, Allolobophora chlorotica, L. rubellus*	Stenersen (1979a)
Parathion (OP), carbofuran (Car)	*E. fetida*	Stenersen (1979b)
Aldicarb (Car), carbaryl (Car), carbofuran (Car), oxamyl (Oxime Car), bromophosoxon (OP), parathion (OP), parathion-methyl (OP), trichloronate (OP)	*E. fetida*	Stenersen (1979c)
Carbofuran (Car), Carbaryl (Car), Malathion (OP), parathion (OP), phorate (OP), N-2596 (OP), dasanit (OP), paraoxon (OP), prostigmine (Car), triorthotolyl phosphate (OP), azak (Car)	*L. terrestris*	Stenersen et al. (1973)
Carbaryl (Car), paraoxon (OP)	*E. andrei, E. fetida, E. veneta*	Stenersen et al. (1992)
Benomyl (Benz), carbenzim (Benz), fuberidazole (Benz), thiabendazole (Benz), 1-(benzimidazol-2-yl)-3-butylurea (Benz)	*L. terrestris*	Stringer and Wright (1976)

[a]Pesticide chemical families: (OP), organo phosphorus compound; (Car), carbamate; (Benz), benzimidazole; (OC), organochlorme.

1973). For example, *Lumbricus terrestris* had a 99% reduction in the ChE activity following treatment with OP insecticides, while carbamate insecticides at the same concentration reduced the ChE activity by only 30% (Steneresen et al. 1973). Organochlorine and benzimidazole compounds were generally reported not to cause reduced ChE activity, although some studies with *L. terrestris* have reported an inhibition due to the metabolic products of benzimidazole (Labrot et al. 1996; Niklas 1979; Scaps et al. 1997; Stringer and Wright 1976).

ChE inhibition has also been observed in *Eisenia*, *Pheretima*, and *Lampito* species. For example, in *E. fetida* a reduction in ChE activity has been observed following exposure to parathion (Stenersen 1979b). Similar reduction of ChE activity was found for *P. posthuma* and *Lampito mauritii* exposed to carbaryl and phosphamidon, respectively (Bharathi and Rao 1985; Dikshith and Gupta 1981).

Few studies have been concerned with ChE depression in worms following exposure to metals, and observations from these studies are less conclusive. Labrot et al. (1996) found strong *in vivo* and *in vitro* reduction of ChE activity in *E. fetida* following exposure to lead and uranium in a filter-paper test. In contrast, Scaps et al. (1997) measured no change in the ChE activity of *E. fetida* when exposed to lead and cadmium using an artificial soil exposure system. Niklas (1979) found no observable effects from copper oxychloride on the *in vitro* ChE activity of *Lumbricus terrestris*.

ChE inhibition in general follows a dose–response relationship with the pollutant (Gupta and Sundararaman 1991; Labrot et al. 1996; Stenersen et al. 1973). For example, Dikshith and Gupta (1981) exposed *P. posthuma* to soil treated with different concentrations of carbaryl and measured a dose-dependent inhibition of AChE activity in both the coelomic fluid and the nervous system.

No clear evidence for a link between ChE depression and mortality, or reproduction, in worms has been observed. For example, the mortality of *E. fetida* following carbofuran exposure was eight times higher than for OP-treated worms, although ChE activity was most severely depressed in OP-exposed earthworms (Stenersen et al. 1973). Similarly, Stringer and Wright (1976) observed no *in vivo* inhibition of ChE in *L. terrestris* treated with benomyl, although mortality was observed after 6 days. Stenersen (1979c) found that the longevity of the depression had an important consequence for the toxicity. Nearly total ChE depression followed by a rapid recovery resulted in little toxicity, whereas a rather lower depression and a slower recovery were more toxic. After *E. fetida* were exposed to pesticides, the earthworm developed a tremor but could survive a strong and long-lasting inhibition of E1 when E2 was not inhibited. In contrast, when E2 was inhibited but E1 was less inhibited, the earthworm showed no symptoms of ChE inhibition. Stenersen et al. (1980a) argued that only those compounds that inhibit both ChEs (E1 and E2) for a certain length of time were lethal to worms (Table 4). Depression of ChE activity may also indirectly cause mortality. For example, Gupta and Sundararaman (1991) found a clear inverse relationship between ChE depression and burrow-

Table 4. Remaining Activity of E1 and E2 as Determined by *In Vitro* Carbaryl Inhibition in *Eisenia fetida*.

Treatment	Remaining activity (%)[a]		Disappearance time (min)
	E1	E2	
Control	100	100	1–2
Carbaryl	0	98	8–15
Aldicarb	5.5	14.3	4
Parathion	16.4	45.9	3–7
Paraoxon	105	32.3	1–2

Source: Stenersen (1979c).
[a]E1 is the difference between the total enzyme activity and E2. E1 is an aldicarb-sensitive enzyme, E2 is less sensitive to aldicarb. Disappearance time is the time required for earthworms to disappear into the soil.

ing activity in earthworms, indicating that although the ChE depression may not cause direct mortality it may render the worm unable to avoid unfavorable conditions.

Variability in the ChE activity of worms is little studied. It is, however, known that it is not always the parent compound that causes the toxic effect, but rather its metabolic breakdown product. In *L. terrestris*, ChE activity was not reduced *in vivo* following carbofuran exposure, while *in vitro* studies showed anti-ChE activity of the breakdown product butyl-isocyanate (Niklas 1979; Stringer and Wright 1976). The relative ChE depression may also be tissue dependent. For example, exposing *Lampito mauritii* to phosphoamidon resulted in a less pronounced ChE inhibition in muscle tissue than in nerve tissue (Bharathi and Rao 1985). Similar tissue differences were observed for *P. posthuma* (Dikshith and Gupta 1981). The level of ChE depression measurable is species dependent. Englestad and Stenersen (1991) and Stenersen et al. (1992) observed that the ChE activities of *L. terrestris*, *E. veneta*, and *E. fetida* were different following exposure to pesticides. The involvement of different ChEs (or levels of these) may explain the differences in toxicity of insecticides to various earthworm species. For example, two ChEs (E1 and E2) were observed in *E. fetida* and only one in *L. terrestris*. Other factors such as season, temperature, nutritional status, and reproductive stage or activity may also influence the ChE activity (Jimenez et al. 1988; McDonald et al. 1990; Rattner and Fairbrother 1991). The duration of ChE activity depression following exposure has only been investigated to a limited extent. Stenersen et al. (1973) observed that ChE depression in earthworms caused by carbamates led to faster reversal (30–40 d) to normal levels than depression levels caused by OPs, which remained depressed for more than 50 d.

V. Other Enzymatic Responses

Many enzyme systems in earthworms have been shown to be affected by chemicals, including cytochrome P-450/mixed-function oxidase (MFO), glutathione transferase (GST), aminolevulinic acid dehydratase (ALAD), peroxidases, catalases, and others, as covered in the following subsections.

A. Cytochrome P-450/Mixed-Function Oxidase (MFO)

One of the most important enzyme systems involved in the oxidation of endogenous and exogenous compounds is the cytochrome P-450-dependent monooxygenase or mixed-function oxidase (MFO) system (Livingstone 1990; Nebert and Gonzalez 1987; Stenersen 1992). This system is a universally distributed enzyme system known to be induced by a range of xenobiotic compounds (Livingstone 1990). In earthworms, good evidence exists for the presence of cytochrome P-450 monooxygenases (Berghout et al. 1989, 1991; Eason et al. 1998). Studies on earthworm cytochrome enzyme induction by chemicals have mainly been concerned with *Lumbricus terrestris* and less so with *Eisenia veneta, Lumbricus rubellus*, and *Aporrectodea caliginosa* (Table 5).

Testing the induction of the P-450 system in earthworms has only been documented for a few chemicals, mainly for enzyme substrates known to induce P-450 in mammals. In *L. terrestris*, Liimatainen and Hänninen (1982) observed an induction of P-450 activity as an induction in the ethoxycoumarin-*o*-dealkylase (ECOD) activity. No induction was observed for ethoxyresorufin-*o*-dealkylase (EROD), ethoxycoumarin-*o*-deethylase, benzphetamine demethylase, or ethylmorphine demethylase activities of the microsomes. This result agreed with later findings by Stenersen (1984) and Eason et al. (1998) because they measured ECOD but no EROD activity. Berghout et al. (1990, 1991) measured no activity for either of these enzymes following exposure to the respective substrates but did observe benzyloxyresorufin-*o*-dealkylase activity in *L. terrestris*. Milligan et al. (1986) did not observe induction of P-450 following exposure to 3-methylcholanthrene (3-MC) and phenobarbitol (PB) in *E. veneta* (Rosa), although these compounds are known to induce the levels of P-450 in vertebrates and in invertebrates other than earthworms. The authors suggested that a very high level of organic matter in the food had resulted in an initial high P-450 level in the earthworms. Feeding the earthworms with food containing a lower organic content, however, did not affect the cytochrome P-450 level in earthworms.

No studies have attempted to establish a dose–response relationship between chemical dose and the P-450 level or to link these responses to higher organizational effect levels. Few studies have investigated the natural variability or persistence of the P-450 response. Eason et al. (1998) observed species differences in the P-450 activity level.

B. Glutathione S-Transferase (GST)

Glutathione *S*-transferase (GST) represents an important family of enzymes named for their role as catalysts for the conjugation of various electrophilic compounds with tripeptide glutathione (Saint-Denis et al. 1996; Stegeman et al.

Table 5. Other Enzymatic Responses in Earthworms in Response to Toxic Compounds.

Enzymatic Responses	Chemical[a]	Species	Reference
Cytochrome (Cyt) P-450/mixed-function oxidases (MFO)			
Cyt P-450/MFO	Alkoxyresorufins, ethoxycoumarin, benzphtamine, ethylmorphine, benzyloxyresorufin, ethoxycoumarin	Lumbricus terrestris	Berghout et al. (1991)
Cyt P-450/MFO	O-Deethylase, ethoxyresorufin-O-deethylase, benzphetamine demethylase, ethylmorphine demethylase	L. terrestris	Berghout et al. (1990)
Cyt P-450/MFO	7-Ethoxyresorufin, 7-pentoxyresorufin, 7-benzyloxyresorufin, 7-ethoxycoumarin	L. rubellus, Apporrectodea caliginosa	Eason et al. (1998)
Cyt P-450/MFO	Ethoxycoumarin, ethoxyresourufin, benzphaetamine, ethylmorphine	L. terrestris	Liimatainen and Hänninen (1982)
Cyt P-450/MFO	3-Methylcholanthrene, phenobarbitol	Eisenia veneta	Milligan et al. (1986)
Cyt P-450/MFO	Ethoxycoumarin, ethoxyresourufin	L. terrestris	Stenersen (1984)
Glutathione S-transferase (GST)			
GST induction	Cd, Pb, Zn	E. fetida	Grelle and Descamps (1998)
GST induction	Aldrin (OC), endosulfan (OC), lindane (OC)	Pheretima posthuma	Hans et al. (1993)

Table 5. (Continued).

Enzymatic Responses	Chemical[a]	Species	Reference
GST induction/inhibition	Induction of GST: 2,4-Dinitrobenzene (CDNB), 1,2-dichloro-4-nitrobenzene (DCNB), ethacrylic acid (ETHA), trans-4-phenyl-3-buten-2-one (TRANS), 4-nitropyridine-N-oxide (NPNO), quintozene, parathion-methyl (OP), hexachlorocyclohexane, atrazine (triazine) Inhibitors of GSH–CDNB conjugation: Phenol red, bromosulfophthalein, eosin yellow, rose bengal, cresolphthalein bilirubin, lithocholic acid (sulfamide), dichlofluanid (sulfamide), tolyfluanid (sulfamide), folpet (phthalimide), captan (phthalimide), flopet (phthalimide), tecnazene (OC), quitozene (OC), atrazine (triazine)	Apporectodea rosea, A. caliginosa, Apporectodea longa, Allolobophora chlorotica, E. fetida, E. hortensis, L. rubellus, L. terrestris	Stenersen and Øien (1981)
GST induction	1-Chloro-2,4-dinitrobenzene, 3,4-dichloro-1-nitrobenzene, sulphobromophtlaein, 1,2-epoxy-3-(p-nitrophenoxy), propane, trans-4-phenyl-butyl-3-en-2-one	E. fetida, L. terrestris, L. rubellus, Apporectodea longa, Apporectodea caliginosa, A. chlorotica	Stenersen et al. (1979)
GST induction	Chloro-2,4-dinitrobenzene, 1,2-dichloro-4-nitrobenzene, ethacrynic acid, 1,2-epoxy-3-[p-nitro-fenoxy]propane	E. fetida	Stokke and Stenersen (1993)

Table 5. (Continued).

Enzymatic Responses	Chemical[a]	Species	Reference
Aminolevulinic acid dehydratase (ALAD)			
ALAD inhibition	Pb	*L. terrestris*	Ireland and Fischer (1978)
Hemoglobin	Pb	*E. fetida*	Migula et al. (1977)
Hemoglobin	Pb/road traffic	*L. terrestris*	Rozen and Mazur (1997)
Peroxidase			
Glutathione peroxidase	Pb, U	*E. fetida*	Labrot et al. (1996)
Lipid peroxidase	Pb, U	*E. fetida*	Labrot et al. (1996)
Catalase	Nitrite	*E. fetida*	Arillo and Melodia (1991)
Catalase	Pb, U	*E. fetida*	Labrot et al. (1996)
Others			
Phosphoesterases	bis[*p*-nitrophenyl]phosphate, *p*-nitrophenylphosphate, *p*-nitrophenol	*L. terrestris*	Park et al. (1993)
Alkaline phosphatase	Levamisole, Zn, phenylalanine, homoarginine, imidazole, tryptophan	*E. andrei*	Park et al. (1998)
Gut enzymes	Malathion (OP)	*Drawida willsi, Oclochaetona surensis, Lampito mauritii*	Patnaik and Dash (1993)
Malate dehydrogenase, phosphoglucomutase, glutamate oxalate transferase	Cd, Pb	*E. fetida*	Scaps et al. (1997)

[a](OC), organochlorine; (OP), organophosphorus compound.

1992). They are probably present in all organisms, are major constituents of animal cells, and form a family of biotransformation enzymes that play an important role in binding or conjugating lipophilic molecules with an electrophilic atom. It has been proposed that GSTs have evolved as a defense system, providing detoxification enzymes, toward products formed by reactions with oxygen (Stenersen et al. 1987). This system has been shown to be affected by various chemicals in many aquatic and terrestrial species (Mayer et al. 1992; Peakall and Walker 1994).

GST activity following chemical exposure has been investigated in species of *Eisenia*, *Lumbricus*, *Allolobophora*, and *Aporrectodea* and in *Pheretima posthuma* and *Eisenia hortensis* (see Table 5). The first authors to show GST induction in earthworms were Stenersen et al. (1979), who measured considerable GST activity in six earthworm species following exposure to 1-chloro-2,4-dinitrobenzene (CDNB), low activity following exposure to 3,4-dichloro-1-nitrobenzene, but no activity with sulphobromophthalein, 1,2-epoxy-3-(*p*-nitrophenoxy)-propane or *trans*-4-phenyl-3-butene-2-one (TRANS) as substrates. Stokke and Stenersen (1993) found no effect on GST activity following exposure to chloro-2,4-dinitrobenzene, 1,2 dichloro-4-nitrobenzene, or ethacrynic acid as an electrophilic substrate, all chemicals known to induce GSTs in mammals. Borgeraas et al. (1996) were able to induce GST activity in *E. fetida* and *E. veneta* using 2-nitrobenzenes and ethacrynic acid cumene hydroperoxide. For compounds of greater environmental concern, Stenersen et al. (1979) found GST induction by methyl-parathion, HCH, atrazine, and quintozene. Using five different soils containing elevated levels of Cd, Zn, Pb, and Cu, Grelle and Descamps (1998) found no change in the GST activity in exposed worms compared to worms from noncontaminated soil. A range of compounds has been shown to inhibit the conjugation between GSH and CDNB in *E. fetida*, thus inhibiting the metabolization of CDNB (Stenersen and Øien 1981) (see Table 5).

It is not known whether the GST response follows a dose–response relationship with chemical exposure (Stegeman et al. 1992). No studies were found linking these responses to higher organizational effect levels, nor studies on the natural variability, and comparative few data on the persistence of GST responses.

Large differences in GST induction may exist between even closely related earthworm species and between different tissues (Stenersen 1992; Stenersen and Øien 1981). For example, Stenersen and Øien (1981) and Stenersen (1979b) observed large species and tissue differences in GST induction by CDNB, 1,2-dichloro-4-nitrobenzene (DCNB), ethacrylic acid (ETHA), TRANS, and 4-nitropyridine-*N*-oxide (NPNO). The activity was the most prominent in the nephridia and crop and least prominent in the chloragenous cells, gizzard, and blood. Time of measurement may also be important. For example, Hans et al. (1993) observed the most enhanced GST activity 1 wk after exposure of *P. posthuma* to aldrin, endosulphan, and lindane. After 2 wk, the GST activity was present only for aldrin and lindane, and there was no activity present after 4 wk. It was not known whether this decline represented a decline in activity of the pesticides after 4 wk or a decline in the sensitivity of the GST after prolonged exposure.

C. Aminolevulinic Acid Dehydratase (ALAD)

Delta-aminolevulinic acid dehydratase (ALAD) is an enzyme found in many tissues, active in the synthesis of hemoglobin by catalyzing the formation of porphobilinogen, a precursor of heme. In other organisms, ALAD has proved to be a highly specific and a relatively simple measure of lead exposure, but not necessarily of toxicity (Mayer et al. 1992). In *Lumbricus terrestris*, Ireland and Fischer (1978) observed a lead-induced reduction of ALAD activity in particular within the chloragocyte cytosol, but also in the body wall and intestine. The ALAD activity was not reduced in isolated chloragosomes. This observation agreed with the reduced hemoglobin contents measured in *Eisenia fetida* exposed to lead and in *L. terrestris* exposed to roadside soil containing high levels of lead (Migula et al. 1977; Rozen and Mazur 1997). No reports were found linking the changes in the ALAD activity, or hemoglobin content, to mortality or reproduction in earthworms. Neither have studies on natural variability and persistence in time been undertaken.

D. Peroxidases

Peroxidases comprise a large array of enzymes that collectively reduce a variety of peroxides to their corresponding alcohol (Frew and Jones 1984; Saint-Denis et al. 1996). *In vivo* exposure of *Eisenia fetida* to lead and uranium did not change the glutathione peroxidase (GPx) activity (activity/mg protein). However, *in vitro* exposure to lead caused a reduced GPx activity and exposure to uranium increased the GPx activity (Labrot et al. 1996). Little is known about the natural variation in the activity of this enzyme, but seasonal changes in GPx activity have been observed in various tissues of the earthworm *Lumbricus terrestris* (Lovas et al. 1987).

The most widely employed assay for lipid peroxidation is the thiobarbituric acid test for malonaldehyde (MDA) (Stegeman et al. 1992). In earthworms, Labrot et al. (1996) observed a clear reduction in MDA levels upon *in vitro* and *in vivo* exposure to lead and uranium. These results agreed with previous results for mammals, fish, and molluscs (Labrot et al. 1996; Stegeman et al. 1992; Sunderman 1986).

E. Catalases

As for peroxidases, catalases are hematin-containing enzymes that facilitate the removal of H_2O_2. They are localized in the peroxisomes of most cells and are involved in fatty acid metabolism (Frew and Jones 1984; Prentø and Prentø 1984). Labrot et al. (1996) observed reduced catalase activity in *Eisenia andrei* upon *in vivo* exposure to lead and cadmium. *In vitro* studies showed similar trends (Labrot et al. 1996). Little is known about the natural variability of the activity of this enzyme, although the activity of catalases in *Lumbricus terrestris* is known to be seasonally dependent (Lovas et al. 1987).

F. Others

A range of other metabolic processes may be used as biomarkers of exposure to chemical compounds. Cancio et al. (1995a,b) measured increased acid phosphatase activity of the chloragogenous tissue of earthworms following exposure to lead. Histologically, they observed a diffusion of the phosphatase activity to the cytosol upon exposure; activity in the control earthworms was confined to the chloragosomes. Patnaik and Dash (1993) studied the effect of malathion on the gut enzymes of *Lampito mauritii*, *Drawida willsi*, and *Oclochaetona surens*. Following exposure, they measured decreased amylase, invertase, cellulase, and urease activities. Scaps et al. (1997) exposed *Eisenia fetida* to cadmium and lead, but observed no effect on the metabolic enzymes malate dehydrogenase, phosphoglucomutase, or glutamate oxalate transferase.

In worm coelomocytes, Ville et al. (1995) found that the lysozyme activity was increased after exposure to Polychlorinated biphenyls (PCBs) in *Eisenia andrei*, *E. hortensis*, and *Lumbricus terrestris* whereas the leukocyte and macrophage viability decreased. Ville et al. (1997) observed that carbaryl and 2,4-dichlorophenoxy acetic acid (2,4-D) reduced the lysozyme activity, cytolytic activity, protease activity, and serpine activity of *E. andrei* coelomic fluid and cytosol.

VI. Energy Reserve Responses

In times of excess food intake which is surplus to that required for maintenance, growth, and reproduction, excess energy is stored in various organs as glycogen or lipid. Such reserves are then mobilized in times of insufficient supply of food to provide energy. Storage of these compounds may be affected by general physiological stressors but can also be influenced by chemical stressors (Mayer et al. 1992). The influences of pollutants on such energy stores have been studied in *Lumbricus rubellus*, *Dendrodrilus rubidus*, *Pheretima posthuma*, *Aporrectodea caliginosa*, and *Eisenia fetida* (Table 6).

The glycogen content of the chloragogenous tissue in *E. fetida* was reduced following exposure to NaF and CaF_2 (Vogel and Seifert 1992). Using the OP insecticides dimethoate and profenofos, El-Banhawy et al. (1984) observed marked depletion of a polysaccharide in the gut epithelial cells of *Ap. caliginosa*. Lead exposure reduced the glycogen level (expressed on the basis of total gram protein) in both the chloragocytes and the intestine of *L. rubellus* (Ireland and Richards 1977). The authors suggested that at concentrations above a certain threshold the energetic cost of storing lead prevented the accumulation of glycogen. The lipid content of *E. fetida* chloragogenous tissue was reduced after exposure to NaF and CaF_2 and other fluorides, as were the lipid stores in the gut epithelial cells of *Ap. caliginosa* following exposure to dimethoate and profenofos (El-Banhawy et al. 1984; Vogel and Seifert 1992).

Few studies have investigated dose–response relationships for the energy reserves compounds. *P. posthuma* exposed to carbaryl showed a dose-dependent

Table 6. Energy Reserve Responses in Earthworms in Response to Toxic Compounds.

Energy reserve response	Chemical[a]	Species	Reference
Glycogen	Carbaryl (Car)	*Pheretima posthuma*	Dikshith and Gupta (1981)
Polysaccharide	Dimethoate (OP), profenofos	*Apporectodea caliginosa*	El-Banhawy et al. (1984)
Glycogen, protein	Pb/Zn mine spoil	*Lumbricus rubellus*	Ireland and Richards (1977)
Glycogen	Pb/Zn mine spoil	*Dendrodilus rubidus*	Richards and Ireland (1978)
Glycogen, lipid	NaF, KF, FCH2-COONa, CaF_2	*Eisenia fetida*	Vogel and Seifert (1992)

[a](Car), carbamate; (OP), organophosphorus compound.

reduction in the clitellar glycogen content (Dikshith and Gupta 1981). No linkage has been observed between the level of energy reserves of earthworms and changes at population levels. Such a link would probably be difficult to demonstrate because organisms under stress may switch their energy consumption from growth to reproduction, or vice versa, depending on the species and environmental conditions. To obtain a linkage to higher levels of effect would require attention to the entire life history of the earthworm.

Little is known with regard to the natural variability in the level of the energy reserves of earthworms, although the levels of such reserves in other organisms have been shown to vary with such factors as food availability, growth stage, reproductive stage, tissue sampled, and ambient temperature. In earthworms, Roots (1960) observed that 2 mon of starvation depleted the glycogen level of the chloragogenous tissue in *L. terrestris*, while this was not observed in *Dendrodrilus rubidus* (Ireland and Richards 1977; Richards and Ireland 1978). It should further be considered that various tissues store glycogen to a different extent (Ireland and Richards 1977).

VII. Lysosomal Membrane Integrity

At the subcellular level, the lysosomal system has been identified as a particular target for the toxic effects of xenobiotics (Moore 1990). Lysosomes are a morphologically heterogeneous group of membrane-bound subcellular organelles, which contain among other substances acid hydrolases, and vary in size up to 1 µm in diameter. The function of lysosomes in the cell is to catabolize organelles and macromolecules (Mayer et al. 1992). Changes in lysosomal membrane stability are thought to be a general measure of stress (Huggett et al. 1992). In stable lysosomes, hydrolases are prevented from reacting with substrates by an intact membrane. The membrane stability decreases in response to stress, and the membrane permeability thus increases. The mechanism(s) causing this alteration is not well understood and may vary with type of stress. Pathological alterations in lysosomes have been especially useful in the identification of adverse environmental impacts on organisms, with much evidence for aquatic organisms but with rather limited evidence for terrestrial organisms (Lowe and Pipe 1994; Lowe et al. 1992; Moore 1985; Svendsen and Weeks 1995; Weeks and Svendsen 1996).

In earthworms, lysosomal integrity changes as a result of chemical stress have been tested in three *Lumbricus* species (*L. terrestris*, *L. castaneus*, and *L. rubellus*) and three *Eisenia* species (*E. andrei*, *E. fetida*, and *E. veneta*). Most studies on lysosomal stability in earthworms have been concerned with the effects of metals, especially copper, and only a few with organic contaminants (Table 7). The first to measure such responses in earthworms were Weeks and Svendsen (1996). In a laboratory study employing *L. rubellus* they observed a significant reduction in the neutral-red retention time (NRR time), a measure of the lysosomal membrane stability of the coelomocytes, with increasing external copper concentrations. A reduced NRR time with increased soil copper concentrations was also observed in a field study with *L. rubellus* (Svendsen and Weeks 1997b), and for *E. veneta* and *E. fetida* exposed to copper and nickel, respectively (Scott-Fordsmand et al. 1998). The NRR times showed a sigmoid dose–response curve when related to the external metal concentrations and clear threshold responses when related to the internal metal levels, above which the NRR times were totally depressed (Svendsen and Weeks 1997a,b; Scott-Fordsmand et al. in press). Studying *L. castaneus* collected at a site polluted by a large fire in a plastic recycling factory, Svendsen et al. (1998) showed a positive correlation between the NRR time and distance from the factory. The NRR time

Table 7. Lysosomal Membrane Integrity Responses (Neutral-Red Retention Time) in Earthworms in Response to Toxic Compounds.

Chemical[a]	Species	Reference
Methiocarb (Car)	*Lumbricus terrestris*	Grafton (1995)
Nickel (Ni)	*Eisenia veneta*	Scott-Fordsmand et al. (1998)
Cu	*E. fetida*	Scott-Fordsmand et al. (in press)
Plastic fire pollution	*L. castaneus*	Svendsen et al. (1996)
Cu	*E. andrei*	Svendsen and Weeks (1997a)
Heavy metals	*L. rubellus*	Svendsen and Weeks (1997b)
Plastic fire	*L. castaneus*	Svendsen et al. (1998)
Cu	*L. rubellus*	Weeks and Svendsen (1996)

[a](Car), carbamate.

showed a dose–response relationship with metal levels present at the site, although the worms at this site also were exposed to organic contaminants (Svendsen et al. 1998). Field exposure of *L. terrestris* to slug bait containing methiocarb showed no effects on the NRR time, although ChE activity was reduced in the worms (Grafton 1995). Using sewage sludge, a reduced NRR time of *E. venta* coelomocytes was observed with increasing amount of sludge (Olesen 1998).

Correlations between the NRR time and reproductive output have been observed for *E. veneta* and *E. fetida* (Scott-Fordsmand et al. 1998). Scott-Fordsmand et al. (in press) showed a correlation between the NRR time and reproduction that appeared to be independent of contamination history. In contrast to this, *E. venta* exposed to increasing amounts of sewage sludge initially gave a reduced NRR time but an increased reproductive output (Olesen 1998). Increasingly more is being learned about the natural variability and persistence in time of lysosomal membrane stability in earthworms.

VIII. Immunological Responses

The immune system is the body's main defense against invasion by foreign material and biological agents (Koller 1993). In recent years, it has been shown that a wide range of chemicals can affect the immune system (Peakall 1994). In mammals, severe immune depression may quickly result in morbidity and death, but sublethal changes in special compartments of the immune system often occur first and provide an early indication of toxic effect. In earthworms, the immune system is located in the coelom, consisting of coelomic fluid and coelomocytes, which like mammalian leukocytes are sensitive to foreign material (Goven et al. 1993b). In earthworms, a variety of different immunological responses have been measured in the nonspecific immune system, specific cell-mediated immune defense, and humoral-mediated defense (Fitzpatrick et al. 1990; Goven et al. 1993a,b; Venables et al. 1992).

The effects of chemicals on the earthworm immune system have been studied in only a few species, especially *Lumbricus terrestris* and to a lesser extent *Eisenia fetida*, *E. andrei*, and *E. hortensis* (Table 8). Most studies have been concerned with the effect of PCBs, although the effects of metals (especially copper), fly ash, and some pesticides have also been investigated (Table 8). For *Lumbricus* species, PCB exposure depressed the ability to form secretory rosettes (S-rosette; S-rosettes are multiple layers of erythrocytes adhering to coelomocytes) and the ability to phagocytize rabbit red blood cells (RRBCs), but not the ability to form E-rosettes (E-rosettes are those coelomocytes with four or more RRBCs adhering to the cell surface in a single layer) (Goven et al. 1987/1988; Rodriguez-Grau et al. 1989). In contrast, Goven et al. (1993a) observed reduced E-rosette formation following exposure to PCB. Maximal depression coincided with maximal internal PCB concentration in the coelomic leukocytes (Rodriguez-Grau et al. 1989). Refuse-derived fuel fly ash depressed

Table 8. Immunological Responses in Earthworms in Response to Toxic Compounds.

Immunological responses	Chemical[a]	Species	Reference
Viability, phagocytosis, agglutination	Cd, polychlorinated phenyl (PCP), phenol, trifluralin	*Eisenia fetida*	Bierkens et al. (1998)
Total immune activity	Polychlorinated biphenyl (PCB), captan (phthalimide), prochloraz (azole), propiconazole (azole), pirimicarb (Car), dimethoate (OP), disulfoton (OP), pirimiphos methyl (OP), paraquat (bipyridylium), chlorpropham (Car), prometon (triazine), tri-allate (Thio-Car)	*Lumbricus terrestris*	Bunn et al. (1996)
Acid phosphatase	Pb	*E. fetida*	Cancio et al. (1995b)
Nitroblue tetrazolium dye reduction	Refuse-derived fuel fly ash	*L. terrestris*	Chen et al. (1991)
Erythrocyte formation, secretory rosettes, phagocytosis	PCB	*L. terrestris*	Fitzpatrick et al. (1990)
Coelomocyte formation, secretory rosettes, erythrocyte rosettes, phagocytosis	PCB	*L. terrestris, E. fetida*	Fitzpatrick et al. (1992)
Phagocytosis	Cd, Hg, Pb, Zn	*L. terrestris*	Fugere et al. (1996)
Phagocytosis	PCP	*L. terrestris*	Giggleman et al. (1998)
Secretory rosettes, erythrocyte rosettes, Phagocytosis	PCB, refuse-derived fuel ash	*L. terrestris*	Goven et al. (1987/1988)

Table 8. (Continued).

Immunological responses	Chemical[a]	Species	Reference
Coelomocyte viability, total extruded cell counts, differential cell counts, erythrocyte formation, secretory rosettes, phagocytosis	PCB	L. terrestris	Goven et al. (1993a)
Coelomocyte viability, total extruded cell counts, differential cell counts, erythrocyte formation, secretory rosettes, phagocytosis	PCB	L. terrestris	Goven et al. (1993b)
Lysozyme activity	Cu	L. terrestris	Goven et al. (1994)
Secretory rosettes	PCB	L. terrestris	Rodriguez-Grau et al. (1989)
Cell viability	PCB	L. terrestris	Suzuki et al. (1995)
Coelomocyte viability, total extruded cell counts, differential cell counts, erythrocyte formation, secretory rosettes, phagocytosis, nitroblue-tetrazolium dye reduction	Hazardous waste sites	L. terrestris, E. fetida	Venables et al. (1992)
Lysozyme activity, leukocytes and macrophage viability	PCB	E. fetida, E. hortensis, L. terrestris	Ville et al. (1995)
Phagocytosis	Carbaryl (Car), 2,4-D (phenoxy)	E. andrei	Ville et al. (1997)
Lysozyme activity	Carbaryl (Car), 2,4-D (phenoxy)	E. andrei	Ville et al. (1997)

[a](Car), carbamate; (OP), organophosphorus compounds.

both the ability to form S-rosettes and the ability to phagocytize, but not the formation of E-rosettes, in *L. terrestris*. Using soil from a hazardous waste site, containing chlordane among other substances, Venables et al. (1992) measured a suppressed spreading of coelomocytes, phagocytosis, and reduction of the dye nitroblue tetrazolium (NBT) in *L. terrestris*. More recently, Bunn et al. (1996) demonstrated a decline in the total immune activity (total immune activity is the combined ability to form rosettes and to phagocytose RRBC) of *L. terrestris* following exposure to pesticides, including carbamates, organophosphorous compounds, azoles, and a few other pesticides (Bunn et al. 1996).

For *Eisenia* species exposed to PCB, responses similar to those for *Lumbricus* species have been observed (Fitzpatrick et al. 1992). The nonspecific immune function, measured by the reduction by coelomocytes of the NBT reduction, was reduced upon exposure to fly ash containing metals (Chen et al. 1991). Acid-washed fly ash (free of metals) did not affect this immune response.

A dose–response relationship between various xenogenic compounds and concomitantly the immunological response has been observed in a number of studies. Rodriguez-Grau et al. (1989) observed a dose–response relationship between PCB dose and the formation of S-rosettes. A similar relationship was observed for the phagocytosis of RRBC, the lysosyme activity, and E-rosettes for worms exposed to carbaryl and 2,4-D (Fitzpatrick et al. 1992; Ville et al. 1995, 1997). For refuse-derived fuel ash (RDF), a dose–response curve was observed for the coelomocyte reduction of NBT and the RDF dose (Chen et al. 1991). Similar results were obtained for the effect of copper on the lysozyme activity in the coelomic fluid and coelomocytes of *L. terrestris* (Goven et al. 1994).

Few studies have been concerned with linking immunological responses to effects at higher organizational levels. Fitzpatrick et al. (1992) observed a correlation between S-rosette formation and mortality in *E. fetida* after exposure to PCB. Immunological responses are known to depend on many factors, but this has been scarcely studied in earthworms (Weeks et al. 1992). The immunological response appears to be species dependent (Fitzpatrick et al. 1992; Diogene et al. 1997), tissue dependent, and dependent on time of sampling (Goven et al. 1994), but there appears to be no or only little seasonal dependency (Eyambe et al. 1991; Goven et al. 1993b, 1994). For example, in copper-exposed *L. terrestris* the immune response was more severely depressed in the coelomic fluid than in the coelomocytes (Goven et al. 1994). *L. terrestris* that were initially exposed to PCB but then transferred to noncontaminated soil showed that their immunologial parameters were restored to preexposure levels (Goven et al. 1993a; Rodriguez-Grau et al. 1989).

IX. Sperm Quality Responses

Changes in the number of sperm and sperm quality in adult individuals have in recent years gained great interest as a marker of pollution in mammals, especially humans (Toppari et al. 1995). Such measurements have attempted to link increased pollution levels with a decline in the reproductive success of humans.

In earthworms, the number and the integrity of sperm have been studied in three instances using *Lumbricus terrestris*, *Eisenia fetida*, and *Eudrilus eigeniae* (Table 9). Cikutovic et al. (1993) observed that in *L. terrestris* exposed to the organochlorine insecticide chlordane or cadmium the number of sperm cells in testis and seminal vesicles decreased. Upon exposure to cadmium, the onset time of the reduction in sperm numbers varied with exposure concentration, but the absolute depression measured was independent of concentration. Reinecke et al. (1995) observed that the ultrastructure of *Eudrilus eigeniae* sperm showed marked abnormalities in the flagella and nuclear regions following the organochlorine dieldrin exposure. Similar results were obtained following the organochlorine exposure of *E. fetida* to various metal salts, with the type of structural damage depending on the metal (Reinecke and Reinecke 1996).

No studies were found linking the chemical effects on earthworm sperm to higher-level changes, and little on the degree of natural variability of earthworm sperm numbers. The number of sperm may depend on time since last mating (Richards and Fleming 1982).

X. Neurological Responses

Measuring the velocity of nerve impulses, in addition to measuring ChE activity, may assess neurotoxic effects. In earthworms such studies have been concerned with changes in velocity of impulses in medial (MGF) and lateral giant nerve fibers (LGF) (Drewes and Lingamneni 1992). The effects of pollutants on these impulses have only been measured in *Eisenia fetida* and *Lumbricus terrestris*.

Changes in neural impulse velocity have mainly been studied in earthworms exposed to organic pollutants (Table 10). For example, Drewes et al. (1984) observed that dieldrin exposure reduced the impulse velocity of MGF in *E. fetida*. A less pronounced picture was obtained following exposure to the fungi-

Table 9. Sperm Quality Responses in Earthworms in Response to Toxic Compounds.

Sperm quality responses	Chemical[a]	Species	Reference
Sperm count	Cd, chlordane (OC)	*Lumbricus terrestris*	Cikutovix et al. (1993)
Sperm ultrastructure	Cd, Mn, Cu, Pb, Zn	*Eisenia fetida*	Reinecke and Reinecke (1996)
Sperm ultrastructure	Dieldrin (OC)	*Eudrilus eugeniae*	Reinecke et al. (1995)

[a](OC), organochlorine.

Table 10. Neurological Responses (Nerve Fiber Conduction Velocity) in Earthworms in Response to Toxic Compounds.

Chemical[a]	Species	Reference
Carbofuran (Car)	*Lumbricus terrestris*	Drewes and Lingamneni (1992)
Dieldrin (OC)	*Eisenia fetida*	Drewes and Vinnig (1984)
Dieldrin (OC), dimethyl-phthalate, fluorene	*E. fetida*	Drewes et al. (1984)
Benomyl (Benz)	*E. fetida*	Drewes et al. (1987)
Carbaryl (Car), Cd, diazinon (OP), dieldrin (OC), dimethyl-phthalate (DMP), fenvalerate (pyrethroid), diazinon (OP)	*E. fetida, L. terrestris*	Drewes et al. (1988)
d-Limonene	*E. fetida*	Karr et al. (1990)

[a](Car) carbamate; (OC), organochlorine; (Benz), benzimidazole; (OP), organophosphorus compound.

cide benomyl, cadmium, carbaryl, DMP, or fluorene (Drewes et al. 1984, 1987, 1988). In *L. terrestris*, Karr et al. (1990) observed a reduction of the neural velocity of MGF and LGF upon direct application of *d*-limonene, the citrus-derived insecticide. Such reduction may show a dose–response relationship as seen, for example, in *E. fetida* exposed to dieldrin, benomyl, and diazinon (Drewes and Vinnig 1984; Drewes et al. 1984, 1987, 1988). In contrast, earthworms exposed to fluorene (alpha-diphenylene-methane) showed reduced a nerve impulse velocity at low concentrations, but less so or not at all at higher concentrations (Drewes et al. 1988).

Information on correlation between the conduction velocity of the nerves and effects on the population level are scarce. For dieldrin a severe (more than 40%) depression, or prolonged exposure, was necessary before mortality occurred (Drewes and Vinnig 1984; Drewes et al. 1984). Little is known about the natural variability and temporal aspects of nervous conduction velocity in earthworms. The velocity of nerve response differs between body parts and between the nerve fibers (Drewes and Vinnig 1984; Karr et al. 1990). Removal of *E. fetida* following 3-hr dieldrin or diazinon exposure, but not after 48-hr exposure, enabled the neural conductivity to return to near normal (preexposure) levels within 72 hr (Drewes and Vinnig 1984; Drewes et al. 1988). For *d*-limonene the reversal of reduced velocity was both concentration and exposure duration dependent (Karr et al. 1990). Severe starvation during postembryonic growth and development has also been observed to cause a depression in nervous conduct similar to that observed for metal-exposed worms (Drewes et al. 1988).

XI. Histopathological Responses

Histopathological markers are tissue lesions that may signal damaging effects resulting from prior or ongoing exposure to one or more toxic agents. In many groups of animals, histopathological changes have been shown to be markers of toxicant exposure, and in certain cases provide precise information on the toxicant causing such lesions, athough in other cases such changes provide no information (Hinton et al. 1992).

A large number of studies have reported general histopathological changes in earthworms such as bruising, swelling, and tissue necrosis (Table 11). Fewer studies have reported more specific histopathological changes. These more specific studies include the earthworm species *Eisenia fetida*, *Dendrodrilus rubidus*, *Lumbricus terrestris*, *L. rubellus*, and *Octolasium transpadanum*.

Histopathological changes have been reported both following exposure to organic and heavy metal pollution. A large number of studies have reported general histopathological changes such as bruising and swelling, especially in the clitellum region, tissue necrosis, and extrusion of coelomic fluid (Antón et al. 1993; Aspöck and An der Län 1963; Cooper and Roch 1992; El-Banhawy et al. 1986; Gupta and Sundaraman 1988; Hans et al. 1990; Karr et al. 1990; Sileo and Gilman 1975; Stenersen 1979a; Stenersen et al. 1975; Stokke and Stenersen 1993). In addition, Zoran et al. (1986) observed teratogenic effects of benomyl in *E. fetida* under segmental regeneration. The effects have mainly been reported following exposure to organic pollutants, but such changes in earthworms have also been observed following exposure to heavy metal pollution (Gupta and Sundaraman 1990). Hence, such changes seem rather nonspecific. General histopathological changes may therefore, until further evidence has been put forward, be considered as nonspecific markers of toxic conditions.

More detailed histopathological changes have also been described. In *Dendrodrilus rubidus*, histological changes were recorded in the chloragocyte cytoplasm following exposure to lead (Richards and Ireland 1978). Chloragosomes from lead-exposed earthworms were irregularly shaped and contained electron-dense flecks associated with their peripheries, and their debris vesicles had various size aggregates of electron-dense particles inside. Nonexposed earthworms had electron-pale debris vesicles containing membrane fragments. The debris vesicles were more numerous in exposed earthworms compared to control earthworms. In fluoride-exposed *E. fetida*, apical flattening of the chloragocytes was observed (Vogel and Seifert 1992).

Following exposure to benomyl and carbofuran, histopathological studies showed that the chloragogenous and coelomic cells became depleted upon exposure, but were renewed when earthworms were removed from the source of pollution (Fischer 1976). Exposure to carbaryl induced swelling of the nuclei in chloragocytes of *P. posthuma* (Gupta and Sundararaman 1988). For *E. fetida*, marked cellular enlargement followed by full depletion of the chloragogenous tissue took place under paraquat exposure (Fischer and Molnar 1992). Dose–response relationships between the chemical exposure level and the histopatho-

Table 11. Histopathological Responses in Earthworms in Response to Toxic Compounds.

Histopathological responses	Chemical[a]	Species	Reference
Swelling, necrosis, loss of movement	Carbaryl (Car)	Not stated	An der Län and Aspöck (1962)
Swelling	Carbofuran (Car)	Eisenia fetida	Antón et al. (1993)
Swelling, necrosis	Carbaryl (Car)	Lumbricus spp.	Aspöck and An der Län (1963)
Wound healing, transplantation	PCB	Lumbricus terrestris	Cooper and Roch (1992)
Golgi and nerve tissue	Dimethoate, profenofos	Apporectodea caliginosa	El-Banhawy et al. (1986)
Chloragocyte cell	Benomyl (Benz), carbofuran (Car)	L. terrestris, Octolasium transpadanum	Fischer (1976)
Intestinal epithelium	Atrazine (triazine), paraquat (bipyridyl)	E. fetida	Fischer (1989)
Chloragocytes cells	Paraquat (bipyridyl), Cr, Sr	E. fetida	Fischer and Molnar (1992)
Coiling and muscle contractions, swelling	Carbofuran (Car)	L. terrestris, E. fetida	Gilman and Vardanis (1974)
Swelling, size, loss of movement	Carbaryl (Car)	Pheretima posthuma	Gupta and Sundararaman (1988)
Loss of movement, necrosis	Cd	P. posthuma	Gupta and Sundaraman (1990)
Swelling, coiling, loss of movements	Aldrin (OC), endosulfan (OC), heptachlor (OC), lindane (OC)	P. posthuma	Hans et al. (1990)
Chloragocyte cells	Pb, Cu, Zn	L. rubellus	Ireland (1978)
Chloragocyte cells	Pb/Zn mine spoil	L. rubellus, Dendrodilus rubidus	Ireland and Richards (1977)
Swelling, autotomy	d-Limonene	E. fetida	Karr et al. (1990)
Chloragogenous cell	Pb/Zn mine spoil	D. rubidus	Richards and Ireland (1978)
Necrosis, loss of movement	Carbofuran (Car)	Not stated	Sileo and Gilman (1975)

Table 11. (Continued).

Histopathological responses	Chemical[a]	Species	Reference
Coiling, loss of movement, swelling	Aldicarb (Car), carbaryl (Car), carbofuran (Car), oxamyl (Car), paraoxon (OP), parathion (OP), gamma-HCH (OC), PHMD	E. fetida, A. caliginosa, Allolobophora chlorotica, L. rubellus	Stenersen (1979a)
Loss of movement, swelling	Carbofuran (Car), carbaryl (Car), malathion (OP), parathion (OP), phorate (OP), N-2596 (OP), fensulfothion (OP), paraoxon (OP), prostigmine (Car), tri-orthotolyl phosphate (OP), azak (Car)	L. terrestris	Stenersen et al. (1973)
Necrosis	Carbofuran		Stenersen et al. (1975)
Swellings	Chloro-2,4-dinitrobenzene, 1,2-dichloro-4-nitrobenzene, ethacrynic acid, 1,2-epoxy-3-[p-nitro-fenoxy]propane	E. andrei	Stokke and Stenersen (1993)
Coiling, swelling	Chlordane (OC), hazardous waste site	L. terrestris, E. fetida	Venables et al. (1992)
Chloragogenous cells	NaF, KF, FCH_2COONa, CaF_2	E. fetida	Vogel and Seifert (1992)
Teratogenic effects	Benomyl (Benz)	E. fetida	Zoran et al. (1986)

[a](Car), carbamate; (Benz), benzimidazole; (OC), organochlorine; (OP), organophosphorus compounds.

logical effects have not been reported. Linkages between histopathological effects and higher organismal effects also have not been reported.

Little is known about the natural variability and temporal aspects of histopathological changes. Fischer and Molnar (1992) observed that hydration, desiccation, cold, hypoxia, and saline load affected the chloragocytes of E. fetida and that the chloragocyte nuclei were enlarged. Fischer and Molnar (1992) further observed that earthworms surviving exposure to paraquat (the *bipyridillium herbicide*) regenerate the chloragogenous tissue that was initially totally depleted.

XII. Behavioral Responses

Behavioral responses in connection with exposure to chemicals enable organisms to avoid intoxication. In earthworms, this phenomenon is observed when a formaldehyde solution is poured onto the soil surface as a means to bring worms to the soil surface when collecting worms in the field (Edwards and Bohlen 1996). Such behavioral responses are only a useful parameter for compounds that are irritants and obviously not for nonirritants or narcotic compounds. Behavioral changes have been reported following exposure to chemicals in a diverse range of earthworm species, with more detailed studies on *Lumbricus terrestris* and *Eisenia fetida* (Table 12).

General behavioural responses may be prominent in connection with exposure to many organic compounds. For example, massive trembling of *Lumbricus* has been observed following exposure to an undisclosed formulation of carbaryl, rendering it impossible for the earthworms to reenter the soil (An der Län and Aspöck 1962). Frequently repeated episodes of whole-body spasms and tight coiling, followed by partial relaxation, may also be seen following exposure to various pesticides or some hazardous waste site soils (Stenersen et al. 1973). However, these behavioral symptoms are difficult to relate to chemical classes because of their general nature.

More detailed investigations of behavioral changes have been performed, for example, by Eijsackers (1987), who observed that *L. rubellus* avoided copper-contaminated soil when given a choice between clean soil and that copper contaminated. Wentsel and Guelta (1988) observed that *L. terrestris* avoided soil containing brass (70% Cu, 30% Zn) at 35 mg/kg but not at 17 mg/kg. Avoidance responses were also observed for *E. fetida* exposed to soil contaminated with KCl and NH_4Cl, but not when exposed to 2-chloro-acteamide (Yeardley et al. 1996). Studying the transport of slug pellets to burrows, Bieri et al. (1989) observed that *L. terrestris* exhibited little transport of slug pellets containing methiocarb (a carbamate insecticide), but extensive transport of metaldehyde-containing pellets, the latter causing increased mortality. Few studies described a dose–response relationship between the behavioral effects and soil dose, but Wentsel and Guelta (1988) observed a decreasing number of worms in soils contaminated with increasing brass powder concentrations. The worms were able to choose a clean soil. Similar observations have been made for soil contaminated with petroleum hydrocarbonates and several pesticides (Slimak 1997; Stephenson et al. 1998).

Little is known as to whether behavioral alterations in general are linked to effects at higher organizational levels. Avoidance has been observed at concentrations at which severe reduction in survival and reproductive occur (Stephen-

Table 12. Behavioral Responses in Earthworms in Response to Toxic Compounds.

Behavioral response	Chemical[a]	Species	Reference
Avoidance	Metaldehyde, methiocarb (Car)	*Lumbricus terrestris*	Bieri et al. (1989)
Locomotion speed	5-Hydroxy-L-tryptophan (5-HTP), DL-*p*-chlorophenylaniline (PCPA)	*L. terrestris*	Burns et al. (1991)
Locomotion speed	Fluoxetine	*L. terrestris*	Burns et al. (1992)
Migration	Benomyl (Benz)	Various species	Mather and Christensen (1998)
Avoidance	Diazinon (OP), carbaryl (Car), safer fruit, vegetable insect killer, lindane (OC), malathion (OP), acephate (OP), diazinon (OP), captan (phthalimide), metaldehyde, chlorothalonil	*L. terrestris*	Slimak (1997)
Avoidance	Petroleum-contaminated soil	*L. terrestris, Eisenia fetida*	Stephenson et al. (1998)
Avoidance	Mancozeb (DithioCar)	*E. fetida*	Vermeulen and Reinecke (1996)
Avoidance	Brass powder	*L. terrestris*	Wentsel and Guelta (1988)
Avoidance	KCl, NH$_4$Cl, 2-chloroacetamide, hazardous waste site	*E. fetida*	Yeardley et al. (1996)

[a](Car), carbamate; (Benz), benzimidazole; (OC), organochlorine; (OP), organophosphorus compounds.

son et al. 1998; Wentsel and Guelta 1987, 1988; Yeardley et al. 1996). The fact that worms avoid a contaminated area in the field site will of course reduce the population in the area but not necessarily cause mortality.

Behavioral responses may differ between species. For example, *E. fetida* have been observed to burrow after direct injection with carbofuran (a systemic carbamate insecticide), while this was not the case for *L. terrestris* (Gilman and Vardanis 1974). Furthermore, *E. fetida* was able to discriminate between carbofuran-treated and nontreated soils whereas *L. terrestris* was not (Gilman and Vardanis 1974).

XIII. Discussion

In the present review, a range of potential biomarkers of toxic compounds were described for earthworms, including biomarkers from the molecular to the organismal level. These have been described in relation to factors of importance for risk assessment, such as the species tested, sensitivity to toxic compounds, dose–response behavior, linkage to population-level effect, and factors important when applying these markers to field samples. The markers included DNA alterations, metal-binding proteins (MT and MBP), cholinesterases and other enzymatic responses, energy reserves, lysosomal integrity, sperm quality, and immunological, neurological, histological, and behavioral responses.

A large number of earthworm species have been used in biomarker studies, but only a few species, notably *Eisenia fetida* and *Lumbricus terrestris*, have received more comprehensive attention. Emphasis has so far mainly been devoted to identification of the mechanisms involved in these responses. Fewer studies have been concerned with identifying possible dose–response relationships, linkages to population-level effects, temporal aspects, and variability in responses. The experiments reported have been performed under very different conditions. For example, some studies have been conducted by means of soil exposure, but others have been performed as filter-paper, aqueous solution, or *in vitro* exposures. This lack of uniformity makes comparison difficult and in many cases gives little indication of such biomarker responses under natural conditions. Few of the biomarker responses have been tested under field or semifield conditions.

A broad range of chemicals have been tested in respect to biomarker responses in earthworms, but a rigid test regime has not been performed for each biomarker. For example, studies on MBPs were, not surprisingly, almost entirely confined to studies involving cadmium exposures, although other metals and organic compounds are also known to induce MBPs. Likewise, lysosomal membrane stability has been tested almost exclusively for metals. A few biomarkers were studied for a broader range of chemicals, such as immunological responses that respond to both metals and organic compounds, although with different sensitivities.

Rigid testing of possible dose–response relationships for the biomarkers has been performed in only a few cases. The ChE activity was reduced with increasing exposure concentrations especially for OP but also for carbamate pesticides. The correlation between the biomarker response and the pollutant may, however, be complex, as illustrated by the lysosomal membrane stability response. For this response, the shape of the dose–response curve clearly depended on whether the response was compared to the external or internal dose of the pollutant, which illustrates the difficulty of assessing dose–response relationship for biomarkers. The biomarker response may not be straightforward, and may only reflect the bioactive fraction, which may or may not increase or decrease in the same way as the external dose of the compound. It is important to stress that it is not necessary for a single biomarker to respond in a dose–response fashion to be useful. Depledge (1994) and Depledge and Fossi (1994) advocated that a range of markers, rather than each biomarker, may be adopted to show a dose–response relationship, and this may well be a useful tool to assess potential risks.

Few of the potential biomarkers measured have established or even attempted to establish linkages to population- or community-level effects. Some evidence does exist; for example, lysosomal membrane stability has been shown to correlate with reproductive output. In contrast, ChE activity showed no evidence of linkages with effects on the population level. For the other potential markers little or no evidence is present for linkages to higher biological levels, mostly because this issue was not addressed.

Natural variability and temporal variation in biomarker responses have in general been little studied. Some evidence does exist showing that factors other than chemicals are of importance for several biomarker responses. The biomarker response may depend on the species employed and the tissue measured, as shown for neural impulse conductivity, ChE activity, and immunological responses. Apart from species variation, temporal aspects of measuring may be important. For example, ChE activity in earthworms was initially depressed by carbamates but recovered 30–40 d after exposure, while such recovery was not observed 50 d after exposure to OP pesticides. Some immunological markers have also been shown to recover to preexposure levels after termination of earthworm exposure to the PCB source.

In conclusion, biomarkers in earthworms in general have not been rigidly tested in respect to parameters important for risk assessment purposes but show some promise in this respect. Changes in DNA may provide a measure of *in situ* genotoxicity in connection with PAH and dioxin contamination and possibly also with other organic compounds. Lysosomal membrane integrity and immunological responses may provide useful tools for risk assessment of *in situ* heavy metal toxicity, although the influence of confounding factors should be further studied. The measurement of earthworm lysosomal membrane integrity may be used as a forewarning of effects at the population level. Esterase activity provides a tool for assessment of exposure of earthworms to pesticides, especially OP and carbamate compounds, but provides little information on the risk to earthworm populations. The time of sampling has been shown to be important for this marker. MBPs provide tools of exposure to certain metals, especially cadmium, but as for the esterase they provide little information on population effects. Behavior appears a useful tool to assess exposure to a diverse range of compounds, but again no population effects can be estimated, and this marker may be very difficult to employ *in situ*. The other biomarkers also show promise as potential tools, but at present too little information relevant to risk assessment is available.

Summary

Earthworms are believed to be so-called key species within ecosystems and are often exposed to a wide range of anthropogenic compounds released to the terrestrial environment. As a consequence, they may suffer from the toxicity of these compounds. For these and other reasons, earthworms have been used extensively in ecotoxicological studies.

In recent years the use of other biological responses (biomarkers) to estimate either exposure or resultant effects of chemicals has received increased attention. Biomarkers address the question of bioavailability by only responding to the bioactive fraction. They may incorporate effects following exposure to a mixture of chemicals. Biomarkers may also reduce extrapolation of results from the laboratory to the field, as they may be applicable under both conditions.

The present review has drawn together current knowledge on potential biomarkers in earthworms and appraised them in relation to basic requirements needed for supplying information relevant to devising satisfactory risk assessment. A wide range of potential biomarkers have been measured in earthworms, including DNA alteration, induction of metal-binding proteins (MTs and MBP), depression of ChE activity and other enzymatic responses, energy reserve responses, responses in neural impulse conductivity, lysosomal membrane stability, immunological responses, changes in sperm numbers, histopathological changes, and behavioral changes. Both organic and inorganic compounds have been included; however, for each biomarker the main emphasis historically has been placed on only a few chemicals. Dose–response relationships were in some cases observed. Little information is available on the linkage of the biomarker response to effects at population or community levels. The influence of other factors, biotic and abiotic, on the biomarker responses and their temporal duration have been only sporadically reported.

Acknowledgment

This review was undertaken as a part of the European BIOPRINT II project, which was partly supported by EU Environmental Research Programme contract no. ENV4-CT96-0222.

References

An der Län von H, Aspöck H (1962) Zur wirkung von sevin auf regenwhürmer. Aus dem Zoologischen Institut der Universität Insbruck, Univ.-Prof. Dr. O. Steinbäck, Vorstand pp 180–182.

Andersen RA, Aune T, Barstad JAB (1978) Characteristics of cholinesterase of the earthworm *Eisenia foetida*. Comp Biochem Physiol C 61:81–87.

Anderson SL, Wild GC (1994) Linking genotoxic responses and reproductive success in ecotoxicology. Environ Health Perspect 102(suppl 12):9–12.

Antón FA, Laborda E, Laborda P, Ramos E (1993) Carbofuran acute toxicity to *Eisenia foetida* Savigny. Earthworms. Bull Environ Contam Toxicol 50:407–412.

Arillo A, Melodia F (1991) Nitrite oxidation in *Eisenia foetida* (Savigny): ecological implications. Funct Ecol 5:629–634.

Aspöck von H, An der Län von H (1963) Ökologische auswirkungen und physiologische besonderheiten des pflanzenschutzmittels sevin (1-naphtyl-*N*-methylcarbamat). Aus dem Zoologischen Institut der Universität Insbruck, Univ.-Prof. Dr. O. Steinbäck, Vorstand, pp 345–380.

Bauer-Hilty A, Dallinger R, Berger B (1989) Isolation and partial characterization of a cadmium-binding protein from *Lumbricus variegatus* (Oligochaeta, Annelida). Comp Biochem Physiol C 94:373–379.

Bengtsson G, Ek H, Rundgren S (1992) Evolutionary response of earthworms to long-term metal exposure. Oikos 63:289–297.

Berghout A, Büld J, Wenzel E (1989) Isolation and partial purification of cytochrome P-450 from the gut of the earthworm *Lumbricus terrestris*. Biol Chem Hoppe-Seyler 370:614.

Berghout A, Büld J, Wenzel E (1990) The cytochrome p450-dependent monooxygenase system of the midgut of the earthworm *Lumbricus terrestris*. Eur J Pharmacol 183: 1885–1886.

Berghout AGRV, Wenzel J, Büld J, Netter KJ (1991) Isolation, partial purification, and characterization of the cytochrome P-450-dependent monooxygenase system from the midgut of the earthworm *Lumbricus terrestris*. Comp Biochem Physiol C 100:389–396.

Bharathi C, Rao BVSS (1985) Inhibition of acetylcholinesterase (AchE) activity by selected organophosphorous pesticides in the earthworm *Lampito-mauritii*. J Environ Biol 6:257–262.

Bieri M, Sweizer H, Christensen K, Daniel O (1989) The effect of metaldehyde and methiocarb slug pellets on *Lumbricus terrestris* Linne. Slugs and snails in world agriculture. BCPC (Br Crop Prot Counc) Monogr 41:237–244.

Bierkens J, Klein G, Corbiseier P, van den Heuvel R, Verschaeve L, Weltens R, Schueters G (1998) Comparative sensitivity of 20 bioassays for soil quality. Chemosphere 37:2935–2947.

Borgeraas J, Nilsen K, Stenersen J (1996) Methods for purification of glutathione transferases in the earthworm genus Eisenia, and their characterization. Comp Biochem Physiol C 114:129–140.

Bremner I, Hoekstra WG, Davies NT, Young BW (1978) Effect of zinc status on rats on the synthesis and degradation of copper-induced metallothioneins. Biochem J 174: 883–892.

Bunn KE, Thompson HM, Tarrant KA (1996) Effect of agrochemicals on the immune systems of earthworms. Bull Environ Contam Toxicol 57:632–639

Burns JT, Bennett II JR, Desmond RE, Meloy JG, Mielke AL, Six JD (1991) A circadian response to 5-HTP inhibition of locomotion in the earthworm, *Lumbricus terrestris*. J Interdiscip Cycle Res 22:255–260

Burns JT, Gryskevich Cl, Artman SA, Hon DM, Huffner JA, Jones BR (1992) Fluoxetine and the circadian rhythm of locomotion in the earthworm, *Lumbricus terrestris*. J Interdiscip Cycle Res 23:218–219.

Cancio I, Gwynn I, Ireland MP, Cajaraville MP (1995a) Lysosomal origin of the chloragosomes in the chloragogenous tissue of the earthworm *Eisenia foetida*: cytochemical demonstrations of acid phosphatase activity. Histochem J 27:591–596.

Cancio I, Gwynn I, Ireland MP, Cajaraville MP (1995b) The effect of sublethal lead exposure on the ultrastructure and on the distribution of acid phosphatase activity in chloragocytes of earthworms (Annelida, Oligochaeta). Histochem J 27:965–973.

Chen SC, Fitzpatrick LC, Goven AJ, Venables BJ, Cooper EL (1991) Nitroblue tetrazolium dye reduction by earthworm (*Lumbricus terrestris*) coelomocytes: an enzyme assay for nonspecific immunotoxicity of xenobiotics. Environ Toxicol Chem 10: 1037–1043.

Cikutovic MA, Fitzpatrick LC, Venables BJ, Goven AJ (1993) Sperm count in earthworms (*Lumbricus terrestris*) as a biomarker for environmental toxicology: effects of cadmium and chlordane. Environ Pollut 81:123–125.

Cooper EL, Roch P (1992) The capacities of earthworms to heal wounds and to destroy allografts are modified by polychlorinated biphenyls (PCB). J Invertebr Pathol 60: 59–63.

Dallinger R (1994) Part B. Detailed report of the contractors. In: Kammenga JE (ed) Progress report 1995 of BIOPRINT. Biochemical fingerprint techniques as versatile tools for the risk assessment of chemicals in terrestrial invertebrates. Third technical report. Report from a workshop held in Innsbruck, Austria, February 9–10, 1996. National Environmental Research Institute, Denmark, pp 21–24.

Dallinger R (1996) Metallothionein research in terrestrial invertebrates: synopsis and perspectives. Comp Biochem Physiol C 113:125–133.

Depledge MH (1994) The rational basis for the use of biomarkers as ecotoxicological tools. In: Fossi MC, Leonzio C (eds) Nondestructive Biomarkers in Vertebrates. Lewis, Boca Raton, pp 271–295.

Depledge MH, Fossi MC (1994) The role of biomarkers in environmental assessment. Invertebrates. Ecotoxicology 3:161–172.

Dikshith TSS, Gupta SK (1981) Carbaryl induced biochemical changes in earthworm (*Pheretima posthuma*). Indian J Biochem Biophys 18:154.

Diogene J, Dufour M, Poirier GG, Nadeau D (1997) Extrusion of earthworm coelomocytes: comparison of the cell populations recovered from the species *Lumbricus terrestris*, *Eisenia fetida* and *Octolasion tyrtaeum*. Lab Anim 31:326–336.

Drewes CD, Lingamneni A (1992) Use of earthworms in eco-neurotoxicity testing: sublethal effects of carbofuran in *Lumbricus terrestris*. In: Grieg-Smith PW, Becker H, Edwards PJ, Heimbach F (eds) Ecotoxicology of Earthworms. Intercept, Hants, UK, pp 197–206.

Drewes CD, Vining EP (1984) In vivo neurotoxic effects of dieldrin on giant nerve fibers and escape reflex function in the earthworm, *Eisenia foetida*. Pestic Biochem Physiol 22:93–103.

Drewes CD, Vining EP, Callahan CA (1984) Environmental toxicology. Non-invasive electrophysiological monitoring: a sensitive method for detecting sublethal neurotoxicity in earthworms. Environ Toxicol Chem 3:599–607.

Drewes CD, Zoran MJ, Callahan CA (1987) Sublethal neurotoxic effects of the fungicide benomyl on earthworms (*Eisenia foetida*). Pestic Sci 19:197–208.

Drewes CD, Vining EP, Callahan CA (1988) Electrophysiological detection of sublethal neurotoxic effects in intact earthworms. In: Edwards CA, Neuhauser EF (eds) Earthworms in Waste and Environmental Management. SPB Academic, The Hague, Netherlands, pp 355–366.

Eason CT, Booth LH, Brennan S, Ataria J (1998) Cytochrome P450 activity in three earthworm species. In: Sheppard S, Bambridge J, Holmstrup M, Posthuma L (eds) Advances in Earthworm Ecotoxicology. SETAC Press, Pensacola, FL, pp 191–198.

EEC (1985) European Economic Community, Directive 79/831, Annex V, Part C: Methods for the determination of ecotoxicity—level 1. DG XI/127-129/82, Rev. 1: Toxicity for earthworms. Commission of the European Community, Brussels.

Edwards CA, Bohlen PJ (1992) The effects of toxic chemicals on earthworms. Rev Environ Contam Toxicol 125:23–100.

Edwards CA, Bohlen PJ (1996) Biology and ecology of earthworms. Chapman & Hall, London.
Edwards CA, Fischer SW (1991) The use of cholinesterase measurement in assessing the impact of pesticides on terrestrial and aquatic invertebrates. In: Mineau P (ed) Cholinesterase-Inhibiting Insecticides—Their Impact on Wildlife and the Environment. Elsevier, Amsterdam, pp 255–275.
Eijsackers H (1987) The impact of heavy metals on terrestrial ecosystems: biological adaptation through behavioural and physiological avoidance. In: Ravera O (ed) Ecological Assessment of Environmental Degradation, Pollution and Recovery. Elsevier, Amsterdam, pp 245–259.
El-Banhawy MA, El-Ganzuri MA, El-Akkad MM (1984) Histochemical studies on polysaccharides and lipids of the gut epithelial cells of the earthworm *Allolobophora caliginosa* living on soil contaminated with insecticides. Biologia 30:173–181.
El-Banhawy MA, El-Ganzuri MA, El-Akkad MM (1986) Morphological changes of the golgi apparatus of the nerve and intestinal cells of the earthworm *Allolobophora caliginosa* living on insecticides-contaminated soil. Pak J Zool 18:1–7.
Engle DW, Brouwer M (1987) Metal regulation and molting in the blue crab, *Callinectes sapidus*. Metallothionein function in metal metabolism. Biol Bull 172:69–81.
Englestad F, Stenersen J (1991) Acetylesterase pattern in the earthworm genus *Eisenia* (*Oligochaeta lumbricidae*)—implications for laboratory use and taxonomic status. Soil Biol Biochem 23:243–247.
Eyambe GS, Goven AJ, Fitzpatrick LC, Venables BJ, Cooper EL (1991) A non-invasive technique for sequential collection of earthworm (*Lumbricus terrestris*) leukocytes during subchronic immunotoxicity studies. Lab Anim 25:61–67.
Fairbairn DW, Olive PL, O'Niel KL (1995) The comet assay: a comprehensive review. Mutat Res 339:37–59.
Fischer E (1976) Chloragocyte-eleocyte transformation induced by benomyl and carbofuran toxication of earthworms (Oligochaeta). Zool Anz (Jena) 197:225–233.
Fischer E (1989) Effects of atrazine and paraquat-containing herbicides on *Eisenia foetida* (Annelida, Oligochaeta). Zool Anz (Jena) 223:291–300.
Fischer E, Molnar L (1992) Environmental aspects of the chloragogenous tissue of earthworms. Soil Biol Biochem 24:1723–1727.
Fitzpatrick LC, Goven AJ, Venables BJ, Rodriguez J, Cooper EL (1990) Earthworm immunoassays for evaluating biological effects of exposure to hazardous materials. In: Sandhu SS (ed) In Situ Evaluations of Biological Hazards of Environmental Pollutants. Plenum Press, New York, pp 119–129.
Fitzpatrick LC, Sassani R., Venables BJ, Goven AJ (1992) Comparative toxicity of polychlorinated biphenyls to earthworms *Eisenia foetida* and *Lumbricus terrestris*. Environ Pollut 77:65–69.
Fossi M, Leonsio C (1994) Non-Destructive Biomarkers in Vertebrates. Lewis, Boca Raton.
Frew JE, Jones P (1984) Structure and functional properties of peroxidases and catalases. In: Sykes AG (ed) Advances in Inorganic and Bioinorganic Mechanisms, Vol. 3. Academic Press, New York, pp 175–212.
Fugere NP, Brousseau K, Krzystyniak D, Coderre, Fournier M (1996) Heavy metal-specific inhibition of phagocytosis and different in vitro sensitivity of heterogenous coelomocytes from *Lumbricus terrestris* (Oligochaeta). Toxicology 109:157–166.
Furst A, Nguyen Q (1989) Cadmium-induced metallothionein in earthworms (*Lumbricus terrestris*). Biol Trace Elem Res 21:81–85.

Gibb JOT, Svendsen C, Weeks JM, Nicholson JK (1997) ^{1}H-NMR spectroscopic investigations of tissue metabolite biomarker response to Cu(II) exposed in terrestrial invertebrates: identification of free histidine as a novel biomarker of exposure to Cu in earthworms. Biomarkers 2:295–302.

Giggleman MA, Fitzpatrick LC, Goven AJ, Venables BJ (1998) Effects of pentachlorophenol on survival of earthworms (*Lumbricus terrestris*) and phagocytosis by their immunoactive coelomocytes. Environ Toxicol Chem 17:2391–2394.

Gilman AP, Vardanis A (1974) Carbofuran comparative toxicity and metabolism in the worms *Lumbricus terrestris* L. and *Eisenia foetida* S. J Agric Food Chem 22:625–628.

Goven AJ, Venables BJ, Fitzpatrick LC, Cooper EL (1987/1988) Original research—an invertebrate model for analyzing effects of environmental xenobiotics on immunity. Clin Ecol v:150–154.

Goven AJ, Eyambe GS, Fitzpatrick LC, Venables BJ, Cooper EL (1993a) Cellular biomarkers for measuring toxicity of xenobiotics: effects of polychlorinated biphenyls on earthworm *Lumbricus terrestris* coelomocytes. Environ Toxicol Chem 12:863–870.

Goven AJ, Fitzpatrick LC, Venables BJ (1993b) Chemical toxicity and host defense in earthworms—an invertebrate model. Ann NY Acad Sci 712:280–300.

Goven AJ, Chen SC, Fitzpatrick LC, Venables BJ (1994) Lysozyme activity in earthworm (*Lumbricus terrestris*) coelomic fluid and coelomocytes: enzyme assay for immunotoxicity of xenobiotics. Environ Toxicol Chem 13:607–613.

Grafton MJ (1995) The effect of application of methiocarb using drilled and broadcast techniques on the earthworm *Lumbricus terrestris*. M.Sc. thesis, Reading University, UK.

Grelle C, Descamps M (1998) Heavy metal accumulation by *Eisenia fetida* and its effect on glutathione-S-transferase activity. Pedobiologia 42:289–297.

Grüber C, Berger B, Gehrig P, Hunziker P, Dallinger R (1998) Characterisation of metallothioneins in the earthworm *Eisenia foetida*. In: Abstracts, 8th Annual Meeting of SETAC–Europe, April 14–18, 1998. Bordeaux, France, p 49.

Gupta SK, Sundararaman V (1988) Carbaryl-induced changes in the earthworm *Pheretima posthuma*. Indian J Exp Biol 26:688–693.

Gupta SK, Sundararaman V (1990) Biological response of earthworm *Pheretima posthuma* of inorganic cadmium. Indian J Exp Biol 28:71–73.

Gupta SK, Sundaraman V (1991) Correlation between burrowing capability and AchE activity in the earthworm, *Pheretima posthuma*, on exposure to carbaryl. Bull Environ Contam Toxicol 46:859–865.

Hans RK, Gupta RC, Beg MU (1990) Toxicity assessment of four insecticides to earthworm, *Pheretima posthuma*. Bull Environ Contam Toxicol 45:358–364.

Hans RK, Khan MA, Farooq M, Beg MU (1993) Glutathione-S-transferase activity in an earthworm (*Pheretima posthuma*) exposed to three insecticides. Soil Biol Biochem 25:509–511.

Hinton DE, Baumann PC, Gardner GR, Hawkins WE, Hendricks JD, Murchelano RA, Okihiro MS (1992) Histopathological biomarkers. In: Huggett RJ, Kimerle RA, Merle PM, Bergman HL (eds) Biomarkers: Biochemical, Physiological, and Histological Markers of Anthropogenic Stress. Lewis, Boca Raton, pp 155–211.

Hopkin SP (1989) Ecophysiology of Metals in Terrestrial Invertebrates. Elsevier, London.

Huggett RJ, Kirmle RA, Mehrle OM, Bergman HL (1992) Biomarkers: Biochemical, Physiological, and Histological Markers of Anthropogenic Stress. Lewis, Boca Raton.

Ireland MP (1978) Heavy metal binding properties of earthworm chloragosomes. Acta Biol Acad Sci Hung 29:385-394.
Ireland MP, Fischer E (1978) Effect of Pb^{++} on Fe^{+++} tissue concentrations and delta-aminolaevulinic acid dehydratase activity in Lumbricus terrestris. Acta Biol Acad Sci Hung 29:395-400.
Ireland MP, Richards KS (1977) The occurrence and localisation of heavy metals and glycogen in the earthworms Lumbricus rubellus and Dendrobaena rubida from a heavy metal site. Histochemistry 51:153-166.
Jimenez BD, Burdis LS, Ezell GH, Egan BZ, Lee NE, McCarthy JF, Beauchamp JJ (1988) Effects of environmental conditions on the mixed function oxidase system in blue gill sunfish (Lepomis macrochirus). Environ Toxicol Chem 7:623-634.
Karr LL, Drewes CD, Coats JR (1990) Toxic effects of d-limone in the earthworm Eisenia fetida (Savigny). Pestic Biochem Physiol 36:175-186.
Klassen CD, Choudhuri S, McKim JM, Lehrman-McKeeman LD, Kershaw WC (1993) Degradation of metallothionein. In: Suzuki KT, Imura N, Kimura M (eds) Metallothionein III. Biological Roles and Medical Implication. Birkhauser, Basel, pp 207-224.
Kokta C (1992) A laboratory test on sublethal effects of pesticides on Eisenia fetida. In: Grieg-Smith PW, Becker H, Edwards PJ, Heimbach F (eds) Ecotoxicology of Earthworms. Intercept, Hants, UK, pp 213-216.
Koller LD (1993) Biomarkers of immunotoxicology. In: Travis CC (ed) Use of Biomarkers in Assessing Health and Environmental Impact of Chemical Pollution. Plenum Press, New York, pp 201-207.
Labrot F, Ribera D, Saint Denis M, Narbonne JF (1996) In vitro and in vivo studies of potential biomarkers of lead and uranium contamination: lipid peroxidation, acetylcholinesterase, catalase and glutathione peroxidase. Biomarkers 1:21-28.
Liimatainen A, Hänninen O (1982) Occurrence of cytochrome P450 in the earthworm Lumbricus terrestris. In: Hietanen E, Laitinen M, Hänninen O (eds) Cytochrome P450: Biochemistry, Biophysics, and Environmental Implication. Proceedings of the 4th International Conference on P-450, Kuopio, Finland, May 31-June 3, Elsevier, Amsterdam, pp 255-257.
Livingstone DR (1990) Cytochrome P-450 and oxidative metabolism in invertebrates. Biochem Soc Trans 18:15-19.
Lloyd-Jones G (1995) ^{32}P-Postlabelling: a valid biomarker for environmental assessment? Toxicol Ecotoxicol News 2:100-105.
Lovas M, Szabo L, Matkovics B, Fischer E (1987) Seasonal antioxidant enzyme changes in earthworm (Lumbricus terrestris) parts. Comp Biochem Physiol C 87:63-64.
Lowe DM, Pipe RK (1994) Contaminant induced lysosomal membrane damage in marine mussel digestive cells: an in vitro study. Aquat Toxicol 30:357-365.
Lowe DM, Moore MN, Evans BM (1992) Contaminant impact on interactions of molecular probes with lysosomes in living hepatocytes from dab Limanda limanda. Mar Ecol Prog Ser 91:135-140.
Mariño F, Stürzenbaum SR, Kille P, Morgan AJ (1998) Cu-Cd interactions in earthworms maintained in laboratory microcosms: the examination of a putative copper paradox. Comp Biochem Physiol C 120:217-223.
Mather JG, Christensen OM (1998) Earthworms surface migration in the field: influence of pesticides using benomyl as test chemical. In: Sheppard S, Bambridge J, Holmstrup M, Posthuma L (eds) Advances in Earthworm Ecotoxicology. SETAC Press, Pensacola, FL, pp 327-340.
Mayer FL, Versteeg DJ, McKee MJ, Folmar LC, Graney RL, McCume DC, Rattner BA

(1992) Physiological and nonspecific biomarkers. In: Huggett RJ, Kimerle RA, Merle PM, Bergman HL (eds) Biomarkers Biochemical, Physiological, and Histological Markers of Anthropogenic Stress. Lewis, Boca Raton, pp 5–86.

McCarthy JF, Shugart LR (1990) Biomarkers of Environmental Contamination. Lewis, Boca Raton.

McDonald IC, Krysan JL, Johnson OA (1990) Studies of electrophoretic variation in diabrotica as influenced by age, sex or diet of adult beetles (Coleptera; Chrysomelidae). Ann Entomol Soc Am 83:1192–1202.

Migula P, Baczkowski G, Wielgus-Serafinska E (1977) Respiratory metabolism and haemoglobin concentration in an earthworm, *Eisenia foetida* (Savigny 1826), under exposure to lead intoxication. Acta Biol (Katowice) 4:79–94.

Milligan DL, Babish JG, Neuhauser EF (1986) Non-inducibility of cytochrome P-450 in the earthworm *Dendrobaena veneta*. Comp Biochem Physiol C 85:85–87.

Min K-S, Itoh N, Okamoto H, Tanaka K (1993) Indirect induction of metallothionien by organic compounds. In: Suzuki KT, Imura N, Kimura M (eds) Metallothionein III. Biological Roles and Medical Implication. Birkhauser, Basel, pp 159–174.

Moore MN (1985) Cellular responses to pollutants. Mar Pollut Bull 16:134–139.

Moore MN (1990) Lysosomal cytochemistry in marine environmental monitoring. Histochemistry 22:187–191.

Morgan JE, Norey CG, Morgan AJ, Kay J (1989) A comparison of the cadmium-binding proteins isolated from the posterior alimentary canal of the earthworms *Dendrodrilus rubidus* and *Lumbricus rubellus*. Comp Biochem Physiol C 92:15–21.

Nebert DW, Gonzalez FJ (1987) P450 genes: structure, evolution, and regulation. Annu Rev Biochem 56:945–993.

Nejmeddine A, Sautiere P, Dhainaut-Courtois N, Baert J-L (1992) Isolation and characterization of a Cd-binding protein from *Allolobophora caliginosa* (Annelida, Oligochaeta): distinction from metallothioneins. Comp Biochem Physiol C 101:601–605.

Niklas von J (1979) Histochemische untersuchungen zur wirkung von pestiziden als cholinesterase-inhibitoren bei *Lumbricus terrestris* L. Z Angew Zool 66:359–368.

OECD (1984) Guideline for testing of chemicals, no. 207. Earthworms: acute toxicity tests. Organization for Economic Cooperation and Development, Paris.

Olesen TME (1998) The use of an earthworm biomarker to assess the effect of sewage sludge applications to agricultural land. M.Sc. report, University of Odense, Denmark.

Park HW, Koh KS, Park SC (1998) Molecular weights and inhibitor sensitivities of alkaline phosphatase isoenzymes from the midgut of the earthworm, *Eisenia andrei*. Soil Biol Biochem 30:831–832.

Park SC, Smith TJ, Bisesi MS (1993) Bioactivation of bis-[*p*-nitrophenyl] phosphate by phosphoesterases of the earthworm, *Lumbricus terrestris*. Drug Chem Toxicol 16: 111–116.

Patnaik HK, Dash MC (1993) Activity of gut enzymes in three tropical grassland earthworm species exposed to sub-lethal malathion suspension. Bull Environ Contam Toxicol 51:780–786.

Peakall DB (1994) Biomarkers. The way forward in environmental assessment. Toxicol Ecotoxicol News 1:55–60.

Peakall DB, Shugart LR (1993) Biomarkers: Research and Application in Assessment of Environmental Health. Springer-Verlag, Berlin.

Peakall DB, Walker CH (1994) The role of biomarkers in environmental assessment. Vertebrates. Ecotoxicology 3:173–179.

Prentø P, Prentø A (1984) Crystalline catalase from the earthworm *Lumbricus terrestris*

(Oligochaeta: Annelida): purification and properties. Comp Biochem Physiol B 77: 325–328.

Ramseier S, Deshusses J, Haerdi W (1990) Cadmium speciation studies in the intestine of *Lumbricus terrestris* by electrophoresis of metal proteins complexes. Mol Cell Biochem 97:137–144.

Rattner BA, Fairbrother A (1991) Biological variability and the influence of stress on cholinesterase activity. In: Mineau P (ed) Cholinesterase-Inhibiting Insecticides: Their Impact on Wildlife and the Environment. Elsevier, Amsterdam, pp 90–107.

Reinecke AJ, Reinecke SA (1996) The influence of heavy metal salts on the growth and reproduction of the compost worms *Eisenia fetida* (Oligochaeta). Pedobiologia 40: 439–448.

Reinecke SA, Reinecke AJ, Froneman ML (1995) The effects of dieldrin on the sperm ultrastructure of the earthworm *Eudrilus eugeniae* (Oligochaeta). Environ Toxicol Chem 14:961–965.

Richards KS, Fleming TP (1982) Spermatozoal phagocytosis by the spermathecae of *Dendrobaena subrubicunda* and other lumbricids (Oligochaeta, Annelida). J Invertebr Reprod 5:233–241.

Richards KS, Ireland MP (1978) Glycogen-lead relationship in the earthworm *Dendrobaena rubida* from a heavy metal site. Histochemistry 56:55–64.

Riddlington JW, Winge DR, Fowler BA (1981) Long-term turnover of cadmium metallothionein in liver and kidney following a single dose of cadmium in rats. Biochem Biophys Acta 673:177–183.

Rodriguez-Grau J, Venables BJ, Fitzpatrick LC, Goven AJ (1989) Suppression of secretory rosette formation by PCBs in *Lumbricus terrestris*: an earthworm assay for humoral immunotoxicity of xenobiotics. Environ Toxicol Chem 8:1201–1207.

Roesijadi G (1993) Response of invertebrate metallotioneins and MT genes to metals and implication for environmental toxicology. In: Suzuki KT, Imura N, Kimura M (eds) Metallothionein, Vol. III. Biological Roles and Medical Implication. Birkhauser, Basel, pp 141–158.

Roots BI (1960) Some observations on the chloragogenous tissue of earthworms. Comp Biochem Physiol 1:218–226.

Rozen A, Mazur L (1997) Influence of different levels of traffic pollution on haemoglobin content in earthworm *Lumbricus terrestris*. Soil Biol Biochem 29:709–711.

Saint-Denis M, Labrot F, Narbonne JF, Ribera D (1996) Glutathione, glutathione-related enzymes, and catalase activities in the earthworm *Eisenia fetida andrei*. Arch Environ Contam Toxicol 35:602–614.

Salagovic J, Gilles J, Verschaeve L, Kalina I (1996) The comet assay for detection of genotoxic damage in earthworms: a promising toll for assessing the biological hazards of polluted sites. Folia Biol (Praha) 42:17–21.

Scaps P, Grelle C, Decamps M (1997) Cadmium and lead accumulation in the earthworm *Eisenia fetida* (Savigny) and its impact on cholinesterase and metabolic pathway enzymes. Comp Biochem Physiol C 116:233–238.

Scott-Fordsmand JJ, Weeks JM, Hopkin SP (1998) Toxicity of nickel to the earthworm and the applicability of the neutral red retention assay. Ecotoxicology 7:291–295.

Scott-Fordsmand JJ, Weeks JM, Hopkins SP (in press) The importance of contamination history for understanding the toxicity of copper to the earthworm Eisenia fetida (Oligochaeta: Annelida) using neutral-red retention assay. Environ. Toxicol. Chem. 000–000.

Shugart L, Theodorakis C (1994) Environmental genotoxicty: probing the underlying mechanisms. Environ Health Perspect 102(suppl 12):13–17.

Shugart L, Bickham J, Jackim G, McMahon G, Ridley W, Stein J, Steinert S (1992) DNA alterations. In: Huggett RJ, Kimerle RA, Merle PM, Bergman HL (eds) Biomarkers: Biochemical, Physiological, and Histological Markers of Anthropogenic Stress. Lewis, Boca Raton, pp 125–154.

Sileo L, Gilman A (1975) Carbonfuran-induced muscle necrosis in the earthworm. J Invertebr Pathol 25:145–148.

Sims RW, Gerard BM (1985) Earthworms. Keys and notes for the identification and study of the species. The Linnean Society of London and The Estuarine and Brackish-Water Science Association, London.

Slimak KM (1997) Avoidance response as a sublethal effect of pesticides on *Lumbricus terrestris* (Oligochaeta). Soil Biol Biochem 29:713–715.

Stegeman JJ, Brouwer M, Di Giulio RT, Forlin L, Fowler BA, Sanders BM, van Veld PA (1992) Molecular responses to environmental contamination: enzyme and protein systems as indicators of chemical exposure and effect. In: Huggett RJ, Kimerle RA, Merle PM, Bergman HL (eds) Biomarkers: Biochemical, Physiological, and Histological Markers of Anthropogenic Stress. Lewis, Boca Raton, pp 235–336.

Stenersen J (1979a) Action of pesticides on earthworms. Part I: The toxicity of cholinesterase-inhibiting insecticides to earthworms as evaluated by laboratory tests. Pestic Sci 10:66–74.

Stenersen J (1979b) Action of pesticides on earthworms. Part II: Elimination of parathion by the earthworm *Eisenia foetida* (Savigny). Pestic Sci 10:104–112.

Stenersen J (1979c) Action of pesticides on earthworms. Part III: Inhibition and reactivation of cholinesterases in *Eisenia foetida* (Savigny) after treatment with cholinesterase-inhibiting insecticides. Pestic Sci 10:113–122.

Stenersen J (1980a) Esterases of earthworms. Part I: Characterisation of the cholinesterases in *Eisenia foetida* (Savigny) by substrates and inhibitors. Comp Biochem Physiol C 66:37–44

Stenersen J (1980b) Esterases of earthworms. Part II: Characterisation of the cholinesterases in *Eisenia foetida* (Savigny) by ion-exchange chromatography and electrophoresis. Comp Biochem Physiol C 66:45–51.

Stenersen J (1984) Detoxification of xenobiotics in earthworms. Comp Biochem Physiol C 78:249–252.

Stenersen J (1992) Uptake and metabolism of xenobiotics by earthworms. In: Grieg-Smith PW, Becker H, Edwards PJ, Heimbach F (eds) Ecotoxicology of Earthworms. Intercept, Hants, UK, pp 129–138.

Stenersen J, Øien N (1981) Glutathione S-transferases in earthworms (*Lumbricidae*). Substrate specificity, tissue and species distribution and molecular weight. Comp Biochem Physiol C 69:243–252.

Stenersen J, Gilman A, Vardanis A (1973) Carbofuran: its toxicity to and metabolism by earthworms (*Lumbricus terrestris*). J Agric Food Chem 21:166–171.

Stenersen J, Gilman A, Vardanis A (1975) Carbofuran-induced muscle necrosis in the earthworm. J Invertebr Pathol 25:145–148.

Stenersen J, Guthenberg C, Mannervik B (1979) Glutathione S-transferases in earthworms (*Lumbricidae*). Biochem J 181:47–50.

Stenersen J, Kobro S, Bjerke M, Arend U (1987) Glutathione transferases in aquatic and terrestrial animals from nine phyla. Comp Biochem Physiol C 86:73–82.

Stenersen J, Brekke E, Engelstad F (1992) Earthworms for toxicity testing: species differ-

ences in response towards cholinesterase inhibiting insecticides. Soil Biol Biochem 24:1761–1764.
Stephenson GL, Kaushik A, Kaushik NK, Solomon KR, Steele T, Scroggins RP (1998) Use of an avoidance response test to assess the toxicity of contaminted soils to earthworms. In: Sheppard S, Bambridge J, Holmstrup M, Posthuma L (eds) Advances in Earthworm ecotoxicology. SETAC Press, Pensacola, FL, pp 67–81.
Stokke K, Stenersen J (1993) Non-inducibility of the glutathione transferases of the earthworm *Eisenia andrei*. Comp Biochem Physiol C 106:753–756.
Stringer A, Wright MA (1976) The toxicity of benomyl and some related 2-substituted benzimidazoles to the earthworm *Lumbricus terrestris*. Pestic Sci 7:459–464.
Stürzenbaum S, Kille P, Morgan AJ (1996) Heavy metal pollution: the earthworm response. In: Abstracts, XII International Colloquium on Soil Ecology, July 21–26, Dublin, Ireland.
Stürzenbaum SR, Kille P, Morgan AJ (1998a) The identification, cloning and characterization of earthworm metallothionein. FEBS Lett 431:437–442.
Stürzenbaum SR, Kille P, Morgan AJ (1998b) Heavy metal-induced molecular responses in the earthworm, *Lumbricus rubellus* genetic fingerprinting by direct differential display. Appl Soil Ecol 9:495–500.
Stürzenbaum SR, Kille P, Morgan AJ (1998c) The identification of new heavy metal responsive biomarkers in the earthworm. In: Sheppard S, Bambridge J, Holmstrup M, Posthuma L (eds) Advances in Earthworm Ecotoxicology. SETAC Press, Pensacola, FL, pp 215–224.
Sunderman FW (1986) Metals and lipid-peroxidation. Acta Pharmacol Toxicol 59:248–255.
Suzuki KT, Yamamura M, Mori T (1980) Cadmium-binding proteins induced in the earthworm. Arch Environ Contam Toxicol 9:415–424.
Suzuki MM, Cooper EL, Eyambe GS, Goven AJ, Fitzpatrick LC, Venables BJ (1995) Polychlorinated biphenyls (PCBs) depress allogeneic natural cytotoxicity by earthworm coelomocytes. Environ Toxicol Chem 14:1697–1700.
Svendsen C, Weeks JM (1995) The use of a lysosome assay for the rapid assessment of cellular stress from copper to the freshwater snail *Viviparus contectus* (Millet). Mar Pollut Bull 31:139–142.
Svendsen C, Weeks JM (1997a) Relevance and applicability of a simple earthworm biomarker of copper exposure. I. Links to ecological effects in a laboratory study with *Eisenia andrei*. Ecotoxicol Environ Saf 36:72–79.
Svendsen C, Weeks JM (1997b) Relevance and applicability of a simple earthworm biomarker of copper exposure. II. Validation and applicability under field conditions in a mesocosm experiment with *Lumbricus rubellus*. Ecotoxicol Environ Saf 36:80–88.
Svendsen C, Meharg AA, Freestone P, Weeks JM (1996) Use of an earthworm lysosomal biomarker for the ecological assessment of pollution from an industrial plastics fire. Appl Soil Ecol 3:99–107.
Svendsen C, Spurgeon DJ, Zvezdelin BM, Weeks JM (1998) Lysosomal membrane permeability and earthworm immune-system activity: field-testing on contaminated land. In: Sheppard S, Bambridge J, Holmstrup M, Posthuma L (eds) Advances in Earthworm Ecotoxicology. SETAC Press, Pensacola, FL, pp 225–232.
Teräväinen H (1969) Ultrastructural distribution of chlineesterase activity in the ventral nerve cord of the earthworm *Lumbricus terrestris*. Histochemie 18:177–190.
Toppari J, Christensen P, Giewercman A, Grandjean P, Guillete LJ, JJgou B, Jensen TK,

Jouannet P, Keiding N, Larsen JC, Leffers H, McLachlan JA, Meyer O, Møller J, Meyts ER-D, Scheike T, Sharpe R, Sumpter J, Skakkebæk NE (1995) Male reproductive health and environmental chemicals with estrogenic effects. Miljøproject 290, Ministry of Environment and Energy, Danish Environment Protection Agency, Copenhagen, Denmark.

van Gestel CAM, van Dis WA, Diren-van Breemen EM, Sparenburg PM (1989a) Development of a standardized reproduction toxicity test with the earthworm species *Eisenia fetida andrei* using copper, pentachlorophenol, and 2,4-dichloroaniline. Ecotoxicol Environ Saf 18:305–312.

van Gestel CAM, van Dis WA, Diren-van Breemen EM, Sparenburg PM, Baerselman R (1989b) Influence of cadmium, copper, and pentachlorophenol on growth and sexual development of *Eisenia fetida* (Oligochaeta; Annelida). Biol Fertil Soils 12:117–121.

van Gestel CAM, Zaal J, Diren-van Breemen EM, Baerselman R (1995) Comparison of two test methods for determining the effects of pesticides on earthworm reproduction. Acta Zool Fenn 196:278–283.

van Schooten FJ, Maas LM, Moonen EJC, Kleinjans JCS, van der Oost R (1995) DNA dosimetry in biological indicator species living on PAH-contaminated soils and sediments. Ecotoxicol Environ Saf 30:171–179.

van Welie RTH, Dijck RGJM, Vermeulen NPE, van Sittert NJ (1992) Mercapturic acids, protein adducts, and DNA adducts as biomarkers of electrophilic chemicals. Crit Rev Toxicol 22:271–306.

Venables BJ, Fitzpatrick LC, Goven AJ (1992) Earthworms as indicators of ecotoxicity. In: Grieg-Smith PW, Becker H, Edwards PJ, Heimbach F (eds) Ecotoxicology of Earthworms. Intercept, Hants, UK, pp 197–206.

Vermeulen LA, Reinecke AJ (1996) Effects of the fungicide mancozeb (zinc-manganese ethylene bis) (dithiocarbamate) on survival, body mass and behaviour of the earthworms *E. fetida* (Oligochaeta). In: Abstracts, XII International Colloquium on Soil Ecology, July 21–26, Dublin, Ireland.

Verschaeve L, Gilles J (1995) Single cell gel electrophoresis assay in the earthworm for the detection of genotoxic compounds in soils. Bull Environ Contam Toxicol 54: 112–119.

Verschaeve L, Gilles J, Schoeters J, van Cleuvenbergen R, de Fr JR (1993) The single cell gel electrophoresis technique or comet test for monitoring dioxin pollution and effects. In: Fiedler H, Frank H, Hutzinger O, Pazzefal W, Riss A, Safe S (eds) Organohalogen Compounds, Vol. II. Federal Environmental Agency, Austria, pp 213–216.

Vigh-Teichmann I, Goslar HG (1968) Enzymehistochemische studien am nervensystem. III. Das verhalten einiger hydrolasen im nervensystem des regenwurms (*Eisenia fetida*). Histochemie 14:352–365.

Ville P, Roch P, Cooper EL, Masson P, Narbonne J-F (1995) PCBs increase molecular-related activities (lysozyme, antibacterial, hemolysis, proteases) but inhibit macrophage-related functions (phagocytosis, wound healing) in earthworms. J Invertebr Pathol 65:217–224.

Ville P, Roch P, Cooper EL, Narbonne JF (1997) Immuno-modulator effects of carbaryl and 2,4-D in the earthworm *Eisenia fetida andrei*. Arch Environ Contam Toxicol 32: 291–297.

Vogel J, Seifert G (1992) Histological changes in the chloragogen tissue of the earthworm *Eisenia fetida* after administration of sublethal concentrations of different fluorides. J Invertebr Pathol 60:192–196.

Walsh P, El Adlouni C, Mukhopadhyay MJ, Viel G, Nadeau D, Poirier GG (1995) ^{32}P-postlabeling determination of DNA-adducts in the earthworm *Lumbricus terrestris* exposed to PAH-contaminated soils. Bull Environ Contam Toxicol 54:654–661.

Walsh P, El Adouni C, Nadeau D, Fournier M, Coderre D, Poirer GG (1997) DNA adducts in earthworms exposed to a contaminated soil. Soil Biol Biochem 29:721–724.

Weeks BA, Anderson DP, DuFour AP, Fairbrother A, Goven AJ, Lahvis GP, Peters G (1992) Immunological biomarkers to assess environmental stress. In: Huggett RJ, Kimerle RA, Merle PM, Bergman HL (eds) Biomarkers: Biochemical, Physiological, and Histological Markers of Anthropogenic Stress. Lewis, Boca Raton, pp 211–234.

Weeks JM (1996) The value of biomarkers for ecological risk assessment: academic toys or legislative tools? Appl Soil Ecol 2:215–216.

Weeks JM, Svendsen C (1996) Neutral red retention by lysosomes from earthworm (*Lumbricus rubellus*) coelomocytes: a simple biomarker of exposure to soil copper. Environ Toxicol Chem 15:1801–1805.

Wentsel RS, Guelta MA (1987) Toxicity of brass powder in soil to the earthworm *Lumbricus terrestris*. Environ Toxicol Chem 6:741–745.

Wentsel RS, Guelta MA (1988) Avoidance of brass powder-contaminated soil by the earthworm, *Lumbricus terrestris*. Environ Toxicol Chem 7:241–243.

Yamamura M, Mori T, Suzuki KT (1981) Metallothionein induced in the earthworm. Experientia (Basel) 37:1187–1189.

Yeardley RB, Lazorchak JM, Gast LC (1996) The potential of an earthworm avoidance test for evaluation of hazardous waste sites. Environ Toxicol Chem 15:1532–1537.

Zoran MJ, Heppner TJ, Drewes CD (1986) Teratogenic effects of the fungicide benomyl on posterior segmental regeneration in the earthworm, *Eisenia fetida*. Pestic Sci 17:641–652.

Manuscript received March 16, 1998; accepted August 21, 1999.

Index

Aldicarb, ChE effects earthworms, 123, 125
Aldrin, lizard bioaccumulation, 46
Alkaline phosphatase, biomarker earthworms, 129
Aluminum, lizard residues, 60
Aminolevulinic acid dehydratase, biomarker earthworms, 129, 131
Aquatic ecosystem modeling, mesocosms, 17
Avoidance, biomarker earthworms, 145
Azinphos-methyl, ChE effects earthworms, 123
Azinphos-methyl, effects lizards, 74

Barium, lizard residues, 60
Behavioral responses, biomarker earthworms, 144
Benzomate, effects lizards, 74
Bioaccumulation, inorganics lizards, 60 ff.
Bioaccumulation, pesticides lizards, 46 ff.
Biomarkers in earthworms, 117 ff.

Cadmium, DNA effects earthworms, 119
Cadmium, lizard residues, 61
Carbaryl, ChE effects earthworms, 123, 125
Carbaryl, effects lizards, 74
Catalases, biomarker earthworms, 129, 131
Ce^{137}, effects lizards, 92
Cesium, lizard residues, 61
ChEs, see Cholinesterases, 122
Chlorfenvinphos, effects lizards, 74
Chlorpyrifos, effects lizards, 74
Cholinesterases, biomarkers earthworms, 122
Chromium, lizard residues, 62
Co^{60}, effects lizards, 98
Cobalt, lizard residues, 62
Coelomocyte viability, biomarker earthworms, 137
Compound 1080, effects lizards, 87

Copper, lizard residues, 63
Cyhexatin, effects lizards, 74
Cytochrome P-450, biomarker earthworms, 126, 127

DAEP, effects lizards, 75
DCD, lizard bioaccumulation, 58
DDD, effects lizards, 75
DDD, lizard bioaccumulation, 46
DDE, effects lizards, 75
DDE, lizard bioaccumulation, 47
DDT, effects lizards, 75
DDT, lizard bioaccumulation, 51
Diazinon, effects lizards, 76
Dicofol, effects lizards, 76
Dicofol, lizard bioaccumulation, 52
Dicofol metabolite, lizard bioaccumulation, 58
Dieldrin, effects lizards, 76
Dieldrin, lizard bioaccumulation, 53
Dioxin, DNA effects earthworms, 119
DNA alterations, biomarker earthworms, 119

Earthworm biomarkers, 117 ff.
Earthworms, behavioral responses biomarkers, 144
Earthworms, ChE effects pesticides, 123
Earthworms, cholinesterase biomarkers, 122
Earthworms, energy reserve biomarkers, 132
Earthworms, histopathological responses biomarkers, 141
Earthworms, immunological responses biomarkers, 136
Earthworms, metallothionein biomarkers, 120, 121
Earthworms, neurological response biomarker, 139, 140
Earthworms, scientific names, 117 ff.
Earthworms, sperm quality biomarker, 138, 139

Index

Earthworms, toxicant DNA alterations, 119
Ecological risk assessment, lizards contaminants, 39 ff.
Ecotoxicological risk assessment, mesocosms, 18
Ecotoxicology, mesocosms outdoor aquatic systems, 1 ff.
Endrin, lizard bioaccumulation, 53
Energy reserves, biomarker earthworms, 132
EPN, effects lizards, 78
Estradiol benzoate, effects lizards, 76
Estrogen, effects lizards, 104
Ethyl parathion, effects lizards, 83
EXAMS, mesocosm exposure modeling system, 14

Fadrozole, effects lizards, 78
Fenitrothion, effects lizards, 79
Fenobucarb, effects lizards, 79
Fentin hydroxide, effects lizards, 79

Glutamate oxalate transferase, biomarker earthworms, 129
Glutathione S-transferase, biomarker earthworms, 126, 127
Glycogen, biomarker earthworms, 133
Gut enzymes, biomarker earthworms, 129

HCB (hexachlorobenzene), lizard bioaccumulation, 54
Hexachlorobenzene (HCB), lizard bioaccumulation, 54
Histopathological responses, biomarker earthworm, 141,

Immunological responses, biomarker earthworms, 135

Lead, lizard residues, 63
Leptophos, effects lizards, 80
Limnocorrals, defined, 3
Lindane, effects lizards, 79
Lindane, lizard bioaccumulation, 54
Littoral enclosures (mesocosms), defined, 3
Lizard contaminant studies, 44 ff.
Lizard pesticide bioaccumulation, 46 ff.

Lizards, as contaminant indicators, 39 ff.
Lizards, contaminant ecological risk assessment, 39 ff.
Lizards, contaminant literature reviews, 44 ff.
Lizards, endangered species listed, 41
Lizards, inorganic contaminants, 72
Lizards, inorganics bioaccumulation, 60 ff.
Lizards, lethal effects contaminants, 73
Lizards, pesticide bioaccumulation, 46 ff.
Lizards, pesticide effects, 74
Lizards, radionuclide effects, 92
Lizards, residue/bioaccumulation data, 43
Lizards, sublethal effects contaminants, 74
Lizards, terrestrial ecosystem components, 41
Lizards, X-irradiation effects, 99
Lysosomal membrane integrity, biomarker earthworms, 133, 134
Lysozyme activity, biomarker earthworms, 137

MAFe, effects lizards, 79
Malate dehydrogenase, biomarker earthworms, 129
Malathion, effects lizards, 80
Manganese, lizard residues, 64
MBCP, effects lizards, 80
Meldrin, effects lizards, 81
MEP, effects lizards, 79
Mercury, lizard residues, 65
Mesocosms, chemical application methods, 13
Mesocosms, defined, 2
Mesocosms, ecological engineering, 1 ff.
Mesocosms, ecosystem modeling, 17
Mesocosms, ecotoxicological risk assessment, 18
Mesocosms, experiment duration, 15
Mesocosms, experimental design/statistics, 12
Mesocosms, importance of shape, 5
Mesocosms, initial composition, 6
Mesocosms, organisms to be tested, 10
Mesocosms, outdoor aquatic systems, 1 ff.
Mesocosms, pesticide testing methods, 13
Mesocosms, place in experimental contexts (diag.), 4
Mesocosms, predictive role, 16

Mesocosms, size *vs* self-sustainability, 3
Metal-binding proteins, biomarkers earthworms, 120, 121
Metallothionein, biomarker earthworms, 120
Methaldehyde, effects lizards, 81
Methiocarb, ChE effects earthworms, 123
Methomyl, effects lizards, 82
Methyl parathion, effects lizards, 82
Methyl parathion, lizard bioaccumulation, 55
Metolcarb, effects lizards, 80
MFO (mixed-function oxidase), biomarker earthworms, 126, 127
Mirex, lizard bioaccumulation, 55
Mitomycin, DNA effects earthworms, 119
Mixed-function oxidase (MFO), biomarker earthworms, 126, 127
Molybdenum, lizard residues, 65
Monodechlorinated dicofol, lizard bioaccumulation, 58
MPMC, effects lizards, 80
MTMC, effects lizards, 80

Natural radiation, effects lizards, 102
Neurological responses, biomarker earthworms, 139, 140
Nickel, lizard residues, 65
Nicotine sulfate, effects lizards, 82

Organochlorine pesticides, lizard bioaccumulation, 46 ff.
Outdoor mesocosms, aquatic systems, 1 ff.
Oxine copper, effects lizards, 83

P-450, biomarker earthworms, 126, 127
PAHs, DNA effects earthworms, 119
Paraoxon, ChE effects earthworms, 125
Parathion (ethyl), effects lizards, 83
Parathion (ethyl), ChE effects earthworms, 123, 125
Parathion (ethyl), lizard bioaccumulation, 58
Pb, ChE effects earthworms, 123
Pelagic bags (mesocosms), defined, 3
Pentachlorophenol, DNA effects earthworms, 119
Pentachlorophenol, effects lizards, 85

Peroxidases, biomarker earthworms, 129, 131
Pesticide bioaccumulation, lizards, 46 ff.
Pesticides, ChE effects earthworms, 123
Pesticides, mesocosm testing methods, 13
Phagocytosis, biomarker earthworms, 137
Phenol, DNA effects earthworms, 119
Phenthoate, effects lizards, 86
Phosphamidon, ChE effects earthworms, 123
Phosphoesterases, biomarker earthworms, 129
Phosphoglucomutase, biomarker earthworms, 129
Polysaccharide, biomarker earthworms, 133
Propoxur, effects lizards, 86
Prothiophos, effects lizards, 86
Pyrethrin, effects lizards, 86

Quinoxaline, effects lizards, 86

Radionuclides, effects lizards, 92, 105
Reptiles, as contaminant indicators, 40
Rotenone, effects lizards, 86
Rubidium, lizard residues, 66

Sediments, selection for mesocosms, 7
SETAC-RESOLVE (1992), 2
Sodium fluoroacetate, effects lizards, 87
Sperm quality, biomarker earthworms, 138, 139
Strontium, lizard residues, 66

TDE, effects lizards, 75
TDE, lizard bioaccumulation, 46
Thermonuclear detonation, effects lizards, 101
Trichlorfon, effects lizards, 90
Trifluralin, DNA effects earthworms, 119

Uranium, ChE effects earthworms, 123

Water, selection for mesocosms, 6

X-irradiation, DNA effects earthworms, 119
X-irradiation, effects lizards, 99
Xylylcarb, effects lizards, 80

Zinc, lizard residues, 77